ROOFING READY RECKONER

METRIC AND IMPERIAL DIMENSIONS FOR
TIMBER ROOFS OF ANY
SPAN AND PITCH

ROOFING READY RECKONER

METRIC AND IMPERIAL DIMENSIONS FOR TIMBER ROOFS OF ANY SPAN AND PITCH

By Ralph Goss

Third edition revised by
Chris N. Mindham

**Blackwell
Science**

© 2001 by the estate of Ralph Goss and
Blackwell Science Ltd

Blackwell Science Ltd
Editorial Offices:
Osney Mead, Oxford OX2 0EL
25 John Street, London WC1N 2BS
23 Ainslie Place, Edinburgh EH3 6AJ
350 Main Street, Malden
  MA 02148 5018, USA
54 University Street, Carlton
  Victoria 3053, Australia
10, rue Casimir Delavigne
  75006 Paris, France

Other Editorial Offices:

Blackwell Wissenschafts-Verlag GmbH
Kurfürstendamm 57
10707 Berlin, Germany

Blackwell Science KK
MG Kodenmacho Building
7–10 Kodenmacho Nihombashi
Chuo-ku, Tokyo 104, Japan

Iowa State University Press
A Blackwell Science Company
2121 S. State Avenue
Ames, Iowa 50014-8300, USA

First published in Great Britain by
Crosby Lockwood & Son Ltd 1948
Second edition imperial/metric 1969
Sixth impression published by Granada Publishing
– Technical Book Division 1980
Reprinted by Blackwell Science 9 times, including
2000
Third edition by Blackwell Science 2001

Set in 9.5/11pt Univers
by DP Photosetting, Aylesbury, Bucks
Printed and bound in Great Britain by
MPG Books Ltd, Bodmin, Cornwall

The Blackwell Science logo is a trade mark of
Blackwell Science Ltd, registered at the United
Kingdom Trade Marks Registry

A catalogue record for this title is available from the
British Library

ISBN 0-632-05765-3

Library of Congress
Cataloging-in-Publication Data

Goss, Ralph.
    Roofing ready reckoner: metric and imperial
dimensions for timber roofs of any span and
pitch/Ralph Goss.—3rd ed./revised by Chris N.
Mindham.
      p. cm.
    ISBN 0-632-05765-3 (alk. paper)
    1. Roofs—Handbooks, manuals, etc.
2. Carpentry—Mathematics.   3. Roofing.
4. Engineering mathematics—Formulae.
I. Mindham, C.N. (Chris N.)   II. Title.

TH2401 .G67   2001
695—dc21   00-140120

DISTRIBUTORS

Marston Book Services Ltd
PO Box 269
Abingdon
Oxon OX14 4YN
(*Orders:* Tel: 01235 465500
              Fax: 01235 465555)

USA
Blackwell Science, Inc.
Commerce Place
350 Main Street
Malden, MA 02148 5018
(*Orders:* Tel: 800 759 6102
                    781 388 8250
            Fax: 781 388 8255)

Canada
  Login Brothers Book Company
  324 Saulteaux Crescent
  Winnipeg, Manitoba R3J 3T2
  (*Orders:* Tel: 204 837-2987
                Fax: 204 837-3116)

Australia
  Blackwell Science Pty Ltd
  54 University Street
  Carlton, Victoria 3053
  (*Orders:* Tel: 03 9347 0300
                Fax: 03 9347 5001)

For further information on
Blackwell Science, visit our website:
www.blackwell-science.com

CONTENTS

1 Introduction   1

2 Roofing terminology   2

3 How to use the ready reckoner   9

4 Metric calculation tables   23

5 Imperial calculation tables   99

6 Wall plate and gable strapping   176

7 Wind bracing   178

8 Roofing metalwork   180

9 Tools and equipment   182

1 INTRODUCTION

The aim of this book is to provide a quick and easily usable reference for those constructing roofs for new buildings and extensions using cut roof or trussed rafter construction. The tables include data for common rafters, hips and valley and can also be used for attic construction. This third edition includes the $2\frac{1}{2}°$ pitch increments used in trussed rafter roof construction for pitches from $17\frac{1}{2}°$–$42\frac{1}{2}°$. The data for steeper pitches and mansard roofs up to $75°$ is retained.

The addition of helpful illustrations and text on wall plate and gable end strapping, wind bracing, truss clips and other roofing metalwork plus valuable information on the tools and equipment to carry out the work, extends the guidance on roof construction.

Although the text is in metric units, imperial comparison is shown in brackets after each dimension with separate imperial worked examples. The all important data tables are retained in both metric and imperial units respecting the continuing popularity of this traditional method of measurement.

2   ROOFING TERMINOLOGY

Wall plate   The 'foundation' of the roof usually 50 × 100 mm wide (2″ × 4″), must be bedded solid, level and straight on the top of the wall, or nailed to the timber framed panel and strapped in place to prevent movement from the structure.

Purlin   Member carrying part load of the long common rafters, traditionally placed at right angles to the rafter but now more commonly fixed vertically.

Pitch   The angle made by the slope of the roof with the horizontal. This may be stated in degrees on the drawing, or may have to be measured by protractor from the drawing, or may have to be calculated by measurement if the new work is to match an existing roof.

Ridge   The timber at the top of the roof where the rafters meet, giving a longitudinal tie to the roof structure, commonly 38 mm ($1\frac{1}{2}''$) thick, and of a depth equal to the top cut on the rafter plus approximately 38 mm ($1\frac{1}{2}''$). This depth will depend upon the pitch of the roof and the tile batten thickness.

Common rafter   The timber running from the ridge, down over the purlin if fitted, over the wall plate, and to the back of the fascia.

Jack rafter   The timber running from the hip rafter down over the purlin if fitted, over the wall plate, and to the back of the fascia.

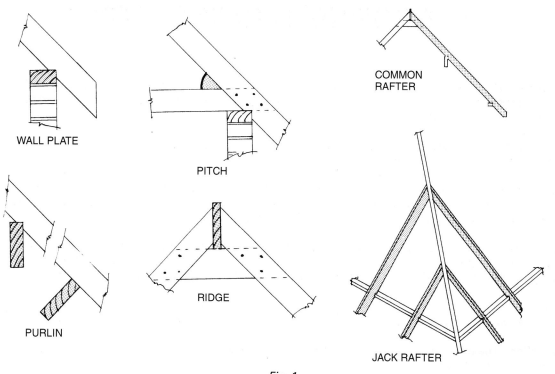

WALL PLATE

PITCH

COMMON
RAFTER

PURLIN

RIDGE

JACK RAFTER

Fig. 1

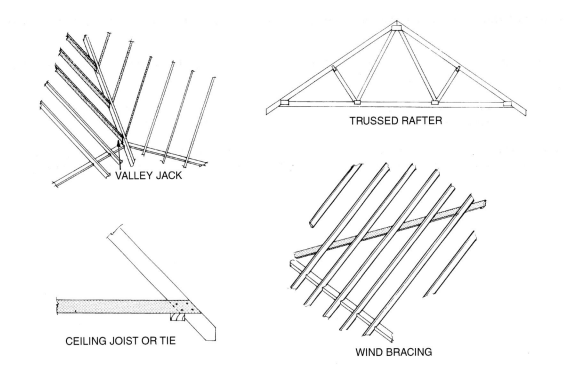

VALLEY JACK

TRUSSED RAFTER

CEILING JOIST OR TIE

WIND BRACING

Fig. 2.

4

Valley jack rafter   The timber running from the ridge, down over the purlin, down to the valley board or rafter.

Trussed rafter   A prefabricated framework incorporating rafter, ceiling joist (or tie), and strengthening webs forming a fully triangulated structural element.

Ceiling joist or tie   Timber supporting the ceiling of the building, but often importantly 'tieing' the feet of the common and jack rafters together thus triangulating and stabilising the roof.

Wind bracing   Usually 25 mm × 100 mm (1″ × 4″) timber nailed to the underside of rafters and trussed rafters running at approximately 45° to them, to triangulate and stabilise the roof in its vertical plain.

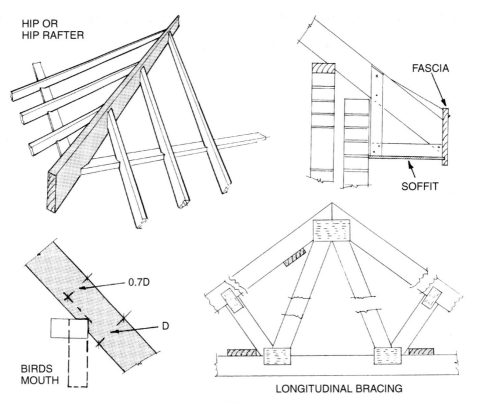

HIP OR
HIP RAFTER

FASCIA

SOFFIT

0.7D

D

BIRDS
MOUTH

LONGITUDINAL BRACING

Fig. 3.

Longitudinal bracing   Usually 25 mm × 100 mm (1″ × 4″) timber nailed to the underside of rafters and trussed rafters both at the ridge position on a trussed rafter roof, and at ceiling joist level on all roofs, to maintain accurate spacing and stiffening of the members to which it is fixed.

Hip or hip rafter   This is a substantial timber member running from the corner of the roof at wall plate level to the end of the ridge. In some designs the hip may stop lower down the roof producing a small gable at high level.

Birdsmouth   The cut in rafters at the fixing point to the wall plate and or the purlin (where purlins are fixed vertically), this should leave at least 0.7 × the depth of the rafter to give the strength necessary for the rafter to continue to provide an over hang to the roof. If a common rafter is fitted as part of a trussed rafter roofing system, then the 0.7 × the depth of the rafter must be the same as the depth of the rafter on the trussed rafter component.

Fascia   Board fixed to the rafter feet, supporting both gutter and soffit.

Soffit   Timber board or sheet material used to close off the over hang between the back of the fascia and the wall. This soffit may have a roof ventilation system built into it.

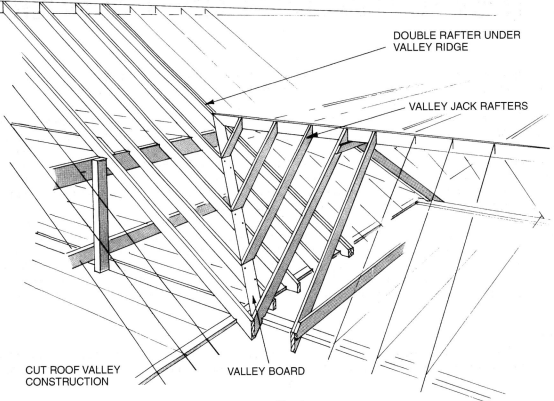

DOUBLE RAFTER UNDER VALLEY RIDGE

VALLEY JACK RAFTERS

CUT ROOF VALLEY CONSTRUCTION

VALLEY BOARD

Fig. 4.

3   HOW TO USE THE READY RECKONER

For the purposes of explaining the use of the ready reckoner, reference should be made to the roof constructions illustrated in Figs 4 and 5. In practical terms these constructions will cover most traditional roof forms and take account of the hip and valley infills used on trussed rafter construction unless fully engineered trussed rafter hip and valleys have been designed. The cutting angles on all timbers for infill rafters on trussed rafter roofs, especially attic designs, can be calculated using the data tables which follow.

Before cutting any of the roof timbers, two vital pieces of information must be known. Firstly the span i.e. the distance between the outer faces of the wall plate, and secondly the 'run' of the rafter, this being half the span assuming that it is an equally pitched roof with the ridge in the middle of the span. Another vital piece of information is the pitch or the 'rise' of the roof.

Whenever possible the carpenter who is to construct the roof should at least supervise the fixing of the wall plates. These must be straight, level and parallel to each other. Where the roof has to be fitted to a 'T' or 'L' plan form of building, then the carpenter should check that the wall plates of the projections to either side of the main roof are at a true right angle unless of course designed to be otherwise. Apart from checking overall dimensions with a steel tape, modern laser levels make it quick and simple to check the level of the plate very effectively, and it is this level of the wall plate which is so important to accurate roof construction.

THE PITCH

The pitch of the roof to be constructed should be clearly stated on the drawings but if not this should be taken by protractor from the drawings, possibly extending the ceiling line and rafter line away from the

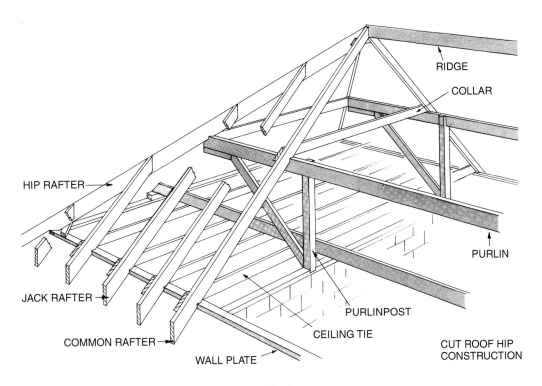

RIDGE

COLLAR

PURLIN

PURLINPOST

CEILING TIE

HIP RAFTER

JACK RAFTER

COMMON RAFTER

WALL PLATE

CUT ROOF HIP
CONSTRUCTION

Fig. 5.

10

point at which they meet making it easier to get an accurate reading from the protractor. An alternative method to establish the pitch is to calculate the rise of the roof per unit of 'run'. To use the tables in this book, this must be stated in metres rise per metre run or in feet and inches rise per foot run. See Fig. 6.

The Run of the Rafter
The run of the rafter is the horizontal distance covered by the rafter from the wall plate to the ridge. See Fig. 7.

The Rise of the Rafter
The rise of the rafter is the height from the top of the rafter vertically above the outside of the wall plate, to the top of the rafter at the centre line of the ridge position. See Fig. 7.

Using the Tables to Cut a Common Rafter
The use of the tables is best explained by a worked example and to do this we will take a roof of pitch at 36° or a rise of 0.727 m per metre run ($8\frac{3}{4}''$ per foot run), and a span of 8.46 m (27' 9"). Then the run:

$$= 8.46 \div 2$$

$$= 4.23 \text{ m } (13' \ 10\frac{1}{2}'')$$

The length of the rafter can now be calculated from the tables referring to 36° pitch. It can be seen that the length of the rafter for 1 m of run = 1.236 m ($1.2\frac{7}{8}$ per 1') therefore the length for 4 m of roof:

$$= 4 \times 1.236$$

$$= 4.944 \text{ m}$$

(for 13', we must add from the tables the 10' and 3' run of rafter giving a total length of $12'4\frac{3}{8}'' + 3'8\frac{1}{2}''$ giving a total of $16'0\frac{7}{8}''$).

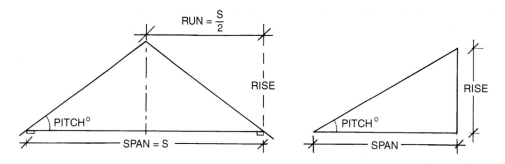

Fig. 6.

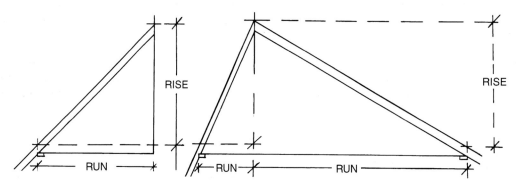

Fig. 7.

12

The calculation for the whole rafter length now looks as follows:

4 m	=	4.944 mm	13'	=	16'0$\frac{7}{8}$"
0.2 m	=	0.247 mm	9"	=	11$\frac{1}{8}$"
0.03 m	=	0.0371 mm	–		–
4.23 m	=	5.2281 mm	13'9"	=	17'0"

Now we have the basic length of the rafter.

To calculate the *exact* length of the rafter, the length above must be reduced by $\frac{1}{2}$ the thickness of the ridge, this can be calculated as above if perfection is required. For a ridge thickness of 40 mm, the length of run of the rafter must be reduced by 20 mm, this gives a reduction in rafter length from the tables of 0.0247 m or 24.7 mm. (Tables have to be modified by a factor of 10 because the run above is 20 mm which is 0.02 m and not 0.2 mm as illustrated in the tables. Similarly on imperial assuming a ridge thickness of 2" which equals a reduction in rafter run of 1", then by reference from the tables it can be seen that the rafter must be reduced by 1$\frac{1}{4}$" in length).

We then need to add to the rafter the extra length needed to cover the over hang. This can be simply calculated in the same way by finding the length of over hang from the outside of the wall plate to the back of the fascia (see Fig. 8), and we will assume for the purposes of this calculation that this over hang gives a 450 mm run (18"), then again by reference to the tables, it will be seen that the additional rafter length required is 512 mm (1' 10$\frac{1}{4}$"). This now gives an overall rafter length as follows:

Basic rafter		5.2281		17'0"
Add over hang	+	0.512	+	1'10$\frac{1}{4}$"
Deduct half ridge	–	0.0247	–	1$\frac{1}{4}$"
		5.7154 mm		18'9"

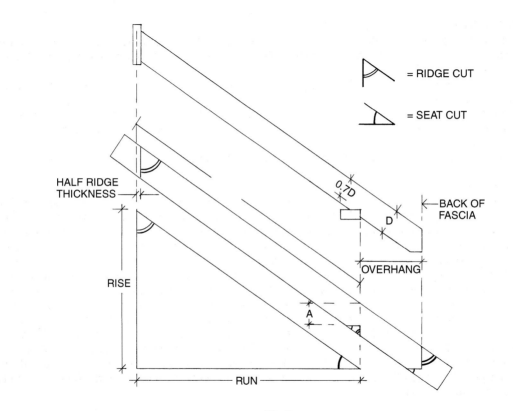

Fig. 8.

DO NOT CUT YET!

Referring now to the tables, mark the ridge bevel on one end for the ridge cut and again at the other end for the fascia cut: in this example the angle will be 54°. The distance between those marks will be the length of the rafter calculated above.

Mark the birdsmouth as shown in Fig. 8 again using the ridge and seat bevels from the tables as indicated in the illustration.

Next mark the soffit line – this must be taken from the building design. If the soffit cut on the rafter is to be at the soffit line, the seat bevel can again be used. If the soffit line is lower than the lowest point of the rafter, then no soffit cut is required.

Now check all dimensions and angles for this first rafter which should be regarded as the master.

NOW CUT THE FIRST RAFTER

Using this first rafter check fit to the roof and use this as a pattern to mark out all of the remaining identical rafters. NB to give a true line for the fascia, it is common practice not to cut the fascia cut on all rafters at this stage. Leave the fascia cut on all roof members until the construction is complete, then using a chalked line from one end of the roof to the other, mark the fascia line on the top of the rafters. From this line, using a level, a true plumb line can be marked and the rafter cut. This traditional method involves cutting the fascia cut on the roof itself which is a time consuming task usually done using a hand saw. Trussed rafter roofs, being prefabricated, generally have the fascia cut made at the factory. If this is the case and allowing for some manufacturing tolerance on span, it will not be possible to line fascia cuts on both sides of the roof. There are two courses of action, (1) to re-mark and cut the foot of the trussed rafter again as outlined above or (2) to use packers to align the fascia onto the pre-cut feet. DO NOT be tempted to align trussed rafters to one side of the roof by aligning their rafter feet. Allowable manufacturing tolerances in a roof of the span we have been discussing could result in a variation of up to 9 mm, thus moving the ridge off the centre line, and up to a 9 mm variance between the feet of the rafters on the opposite side of the roof.

Cutting the common rafter can obviously be done by hand saw, by powered hand saw, or by using a powered compound mitre saw which can be pre-set at the ridge bevel, and then with a saw table with the length stop at an appropriate position, all cuts will be precisely the same with no further marking required.

HIP JACK RAFTERS

The length of these members will depend upon the centres at which they are to be fixed; by that we mean their spacing centre line to centre line of the thickness of the member, which should match the common rafter spacing. See Figs 1 and 9.

Continuing with the example above, the basic common rafter length was 5.2281 m (17'0''), and then assuming a jack rafter spacing of 600 mm (24'') by referring to the table, it can be seen that this length must be reduced by 742 mm $2'5\frac{5}{8}''$.

DO NOT forget to add the over hang of the common rafter; DO NOT adjust the jack rafter for the hip until a trial fit has been taken. The jack rafter meets the hip at an angle and must therefore be cut at an angle to meet the hip both horizontally and vertically, giving what is known as a compound cut. The hip, being fixed vertically in its section, gives the same bevel cut at the top of the jack rafter as was used at the ridge and this same ridge bevel can be used. However, the edge cut can be found in the tables as the 'edge bevel', and for this an example can be seen as 39°. With these two angles the top of the hip jack can be marked, and at the lower end, the common rafter master can be used to mark the fascia cut and birdsmouth.

Cutting compound bevels by hand is a skilled task, but the powered compound mitre saw can be used to produce accurate repeatable compound cuts.

HIP RAFTER

A full hip (i.e. that which is constructed from wall plate to ridge), will have the same rise as the common rafters and the tables have been calculated on the assumption that the hip is on the mitre of a right

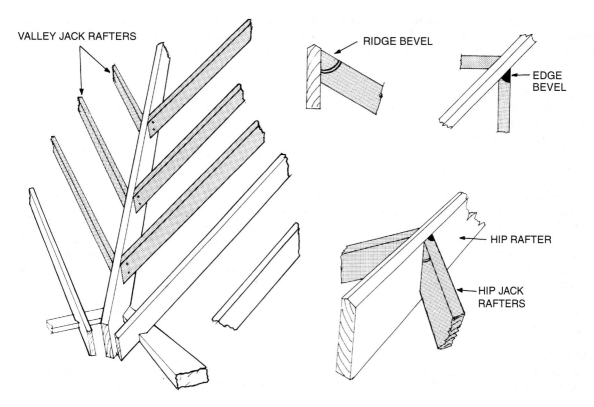

VALLEY JACK RAFTERS

RIDGE BEVEL

EDGE BEVEL

HIP RAFTER

HIP JACK RAFTERS

Fig. 9.

angled corner and is therefore at right angles to the common rafters. This allows the same run as the common rafter to be used to save further calculation.

Continuing with the example above then, the run of the hip would be 4.23 m ($13'10\frac{1}{2}''$), and by referring to the tables for the length of hip, the calculation will result in a hip length of 6.7257 m ($22'0\frac{5}{8}''$).

The seat and ridge bevels can be taken directly from the tables in this case 27° and 63° respectively. Care must be taken when setting out the birdsmouth to ensure that the depth of rafter (A) in the illustration Fig. 10, equals that of the common rafter illustrated in Fig. 8.

The mitre at the top of the hips where they meet the ridge does need the special setting of a bevel. The marking gauge is set to $\frac{1}{2}$ the thickness of the hip and marked on the end of both faces after the plumb cut is made. See Fig. 10.

Backing of Hips

In a good job, the hips are backed – that is to say a chamfer is planed both ways from the centre line on the top of the hip so that the two surfaces are in line with the planes of the roof on adjacent sides. This gives a good seating for the battens. After cutting the hip to the plumb line the same plumb bevel for marking the profile of the backing chamfers on each side of the hip may be used. See Fig. 10.

Dimension B, the length of the plumb cut of the jack rafters is measured on the top end of the hip down from the backing levels leaving a remainder C. If C is measured along the side bevel of purlin it gives the position of the projection under the hip. See Fig. 11.

VALLEY JACK RAFTERS

The tables are based on a construction which assumes a valley rafter similar to the hip rafter, see Fig. 9, NOT that illustrated in Fig. 4 which is a more modern construction and one which works with a trussed rafter roof if a prefabricated valley set of frames is not provided. Returning then to the traditional cut valley, this is essentially a hip in reverse. The valley jacks decrease in length as they progress up the roof, and again would be based at similar centres to the common rafters. The same bevels as for the hip jack

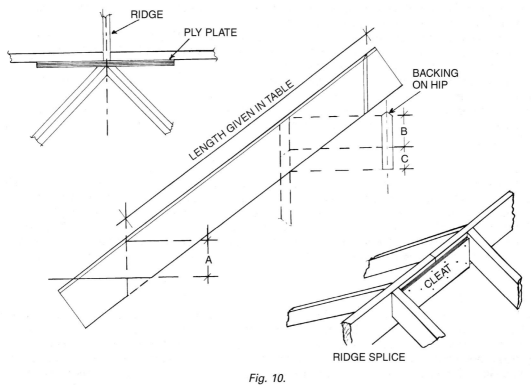

RIDGE

PLY PLATE

BACKING ON HIP

LENGTH GIVEN IN TABLE

B

C

A

CLEAT

RIDGE SPLICE

Fig. 10.

19

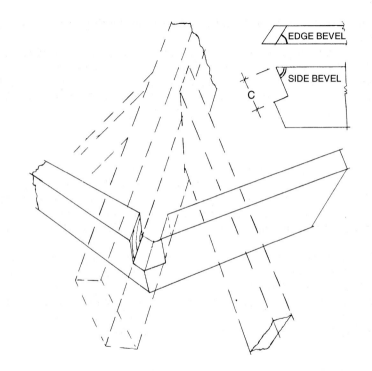

EDGE BEVEL

SIDE BEVEL

C

Fig. 11.

rafter can be used but this time on the foot of the rafter rather than the ridge as before. The common rafter ridge bevel can be used at the top.

Now referring to Fig. 4, the top cut on the valley jack is the same as the common rafter, but the foot of the valley rests on a flat valley board nailed on the top of the rafters of the main roof. This construction is suitable for all valleys except attic construction. The cut at the bottom of the valley jack is the same as the seat cut for the common rafter with an edge bevel equal to the pitch of the main roof. This may not be the same as the pitch of the roof on which the valley jack has to be fitted and should therefore be checked.

THE RIDGE

This roof member is usually relatively thin and can be no more than 25–38 mm (1″–1½″). It takes little load from the roof as pairs of common rafters or valley rafters are placed directly opposite one another across the ridge, it does act as a tie from gable to gable or hip to hip. In a roof with a gable at both ends, the length of the ridge equals the length of the wall plate and is usually built into the gables at both ends. In a hip roof with a hip at both ends, the ridge length is normally the length of the building (internally) less the width (internally). If a short packing piece is used at the end of the ridge in a hip construction to give good hip rafter supports, then this thickness must be deducted from the overall length of the ridge for each hip. The packing piece should be at least as deep as the hip rafter ridge cut, and may therefore be too deep for readily available softwood. The use of 19 mm (¾″) exterior grade plywood is recommended. The ply can also be used to couple ridge pieces together in their length because even the longest lengths of timber may be too short for the overall length of the ridge itself. See Fig. 10. These 'cleats' should be located between rafters to avoid having to cut back the rafter by the thickness of the cleat.

PURLINS

As illustrated under the definition for purlin earlier in this book, Fig. 1, the purlin can be fitted either at right angles to the underside of the rafter or vertically. The tables given in this book show edge and side bevels for purlins set at right angles to the rafters. See Fig. 11. This is the traditional form of construction and may have to be followed on extension work to older buildings.

The purlin fitted vertically is preferred from a structural viewpoint because it acts as a true beam carrying the rafters. The rafters can be better fitted by birdsmouth to the purlin. The edge bevel on this type of purlin is that formed by the hip on plan which on a right angled hip is 45°. The side bevel is 90°, simply a square cut.

4 METRIC CALCULATION TABLES

RISE OF COMMON RAFTER   0.087 m PER METRE OF RUN   PITCH   5°

BEVELS: COMMON RAFTER   – SEAT   5
                    "   "   – RIDGE 85
              HIP OR VALLEY   – SEAT
              "   "   "   – RIDGE
              JACK RAFTER   – EDGE
              PURLIN   – EDGE
              "   – SIDE

JACK RAFTERS   333 mm CENTRES DECREASE   (in mm to 999 and
                  400 "   "   "   thereafter in m)
                  500 "   "   "
                  600 "   "   "

Run of Rafter	0.1	0.2	0.3	0.4	0.5	0.6	0.7	0.8	0.9	1.0
Length of Rafter	0.1	0.201	0.301	0.401	0.502	0.602	0.703	0.803	0.903	1.004
Length of Hip										

RISE OF COMMON RAFTER 0.105 m **PER METRE OF RUN**   **PITCH** 6°

BEVELS: COMMON RAFTER   – SEAT   6
               "   "   – RIDGE 84
            HIP OR VALLEY   – SEAT
               "   "   "   – RIDGE
            JACK RAFTER   – EDGE
            PURLIN   – EDGE
               "   – SIDE

JACK RAFTERS   333 mm   **CENTRES DECREASE**   (in mm to 999 and
                    400   "   "   "   thereafter in m)
                    500   "   "   "
                    600   "   "   "

Run of Rafter	0.1	0.2	0.3	0.4	0.5	0.6	0.7	0.8	0.9	1.0
Length of Rafter	0.101	0.201	0.302	0.402	0.503	0.603	0.704	0.804	0.905	1.006
Length of Hip										

RISE OF COMMON RAFTER 0.123 m **PER METRE OF RUN**   **PITCH** 7°

BEVELS: COMMON RAFTER   – SEAT 7
                            ″   ″   – RIDGE 83
                   HIP OR VALLEY   – SEAT
                   ″   ″   ″   – RIDGE
                   JACK RAFTER   – EDGE
                   PURLIN   – EDGE
                      ″   – SIDE

JACK RAFTERS 333 mm **CENTRES DECREASE**   (in mm to 999 and
                 400 ″   ″   ″   thereafter in m)
                 500 ″   ″   ″
                 600 ″   ″   ″

Run of Rafter	0.1	0.2	0.3	0.4	0.5	0.6	0.7	0.8	0.9	1.0
Length of Rafter	0.101	0.202	0.302	0.403	0.504	0.605	0.705	0.806	0.907	1.008
Length of Hip										

RISE OF COMMON RAFTER 0.141 m PER METRE OF RUN   PITCH 8°

BEVELS: COMMON RAFTER   – SEAT 8
                                ″   ″   – RIDGE 82
                 HIP OR VALLEY   – SEAT
                  ″   ″   ″   – RIDGE
                 JACK RAFTER   – EDGE
                 PURLIN   – EDGE
                      ″   – SIDE

JACK RAFTERS 333 mm CENTRES DECREASE   (in mm to 999 and

                 400 ″   ″   ″   thereafter in m)
                 500 ″   ″   ″
                 600 ″   ″   ″

Run of Rafter	0.1	0.2	0.3	0.4	0.5	0.6	0.7	0.8	0.9	1.0
Length of Rafter	0.101	0.202	0.303	0.404	0.505	0.606	0.707	0.808	0.909	1.01
Length of Hip										

RISE OF COMMON RAFTER 0.158 m **PER METRE OF RUN**   **PITCH** 9°

BEVELS: COMMON RAFTER   – SEAT 9
         "   "   – RIDGE 81
         HIP OR VALLEY   – SEAT
         "   "   "   – RIDGE
         JACK RAFTER   – EDGE
         PURLIN   – EDGE
         "   – SIDE

JACK RAFTERS 333 mm **CENTRES DECREASE**   (in mm to 999 and
                  400 "   "   "   thereafter in m)
                  500 "   "   "
                  600 "   "   "

Run of Rafter	0.1	0.2	0.3	0.4	0.5	0.6	0.7	0.8	0.9	1.0
Length of Rafter	0.101	0.203	0.304	0.405	0.506	0.608	0.709	0.81	0.911	1.013
Length of Hip										

RISE OF COMMON RAFTER 0.176 m **PER METRE OF RUN**   **PITCH** 10°

BEVELS: COMMON RAFTER   – SEAT   10
               "   "   – RIDGE 80
            HIP OR VALLEY   – SEAT
              "   "   "   – RIDGE
            JACK RAFTER   – EDGE
            PURLIN   – EDGE
              "   – SIDE

JACK RAFTERS   333 mm **CENTRES DECREASE**   (in mm to 999 and
                400   "   "   "   thereafter in m)
                500   "   "   "
                600   "   "   "

Run of Rafter	0.1	0.2	0.3	0.4	0.5	0.6	0.7	0.8	0.9	1.0
Length of Rafter	0.102	0.203	0.305	0.406	0.508	0.609	0.711	0.812	0.914	1.015
Length of Hip										

RISE OF COMMON RAFTER 0.194 m **PER METRE OF RUN**   **PITCH** 11°

BEVELS: COMMON RAFTER – SEAT 11
           "   "   – RIDGE 79
        HIP OR VALLEY – SEAT
       "   "   "   – RIDGE
        JACK RAFTER – EDGE
        PURLIN – EDGE
        "   – SIDE

JACK RAFTERS 333 mm **CENTRES DECREASE**   (in mm to 999 and
                400 "   "   "   thereafter in m)
                500 "   "   "
                600 "   "   "

Run of Rafter	0.1	0.2	0.3	0.4	0.5	0.6	0.7	0.8	0.9	1.0
Length of Rafter	0.102	0.204	0.306	0.407	0.509	0.611	0.713	0.815	0.917	1.019
Length of Hip										

RISE OF COMMON RAFTER 0.213 m **PER METRE OF RUN**   **PITCH** 12°

BEVELS:	COMMON RAFTER	– SEAT	12
	″ ″	– RIDGE	78
	HIP OR VALLEY	– SEAT	
	″ ″ ″	– RIDGE	
	JACK RAFTER	– EDGE	
	PURLIN	– EDGE	
	″	– SIDE	

JACK RAFTERS 333 mm **CENTRES DECREASE**   (in mm to 999 and
                     400 ″   ″   ″   thereafter in m)
                     500 ″   ″   ″
                     600 ″   ″   ″

Run of Rafter	0.1	0.2	0.3	0.4	0.5	0.6	0.7	0.8	0.9	1.0
Length of Rafter	0.102	0.204	0.307	0.409	0.511	0.613	0.716	0.818	0.92	1.022
Length of Hip										

RISE OF COMMON RAFTER 0.231 m **PER METRE OF RUN**   **PITCH** 13°

BEVELS: COMMON RAFTER   – SEAT   13
              "   "   – RIDGE   77
           HIP OR VALLEY   – SEAT
           "   "   "   – RIDGE
           JACK RAFTER   – EDGE
           PURLIN   – EDGE
           "   – SIDE

JACK RAFTERS 333 mm **CENTRES DECREASE**   (in mm to 999 and
              400   "   "   "   thereafter in m)
              500   "   "   "
              600   "   "   "

Run of Rafter	0.1	0.2	0.3	0.4	0.5	0.6	0.7	0.8	0.9	1.0
Length of Rafter	0.103	0.205	0.308	0.411	0.513	0.616	0.718	0.821	0.924	1.026
Length of Hip										

BEVELS: COMMON RAFTER   – SEAT 14
               ″   ″   – RIDGE 76
           HIP OR VALLEY   – SEAT
               ″   ″   ″   – RIDGE
           JACK RAFTER   – EDGE
           PURLIN   – EDGE
               ″   – SIDE

JACK RAFTERS 333 mm **CENTRES DECREASE**   (in mm to 999 and
                  400 ″   ″   ″   thereafter in m)
                  500 ″   ″   ″
                  600 ″   ″   ″

Run of Rafter	0.1	0.2	0.3	0.4	0.5	0.6	0.7	0.8	0.9	1.0
Length of Rafter	0.103	0.206	0.309	0.412	0.515	0.618	0.721	0.824	0.928	1.031
Length of Hip										

RISE OF COMMON RAFTER 0.268 m **PER METRE OF RUN**   **PITCH** 15°

BEVELS: COMMON RAFTER   – SEAT   15
                ''   ''   – RIDGE 75
          HIP OR VALLEY   – SEAT
            ''   ''   ''   – RIDGE
          JACK RAFTER   – EDGE
          PURLIN   – EDGE
             ''   – SIDE

JACK RAFTERS 333 mm **CENTRES DECREASE**   (in mm to 999 and
               400 ''   ''   ''   thereafter in m)
               500 ''   ''   ''
               600 ''   ''   ''

Run of Rafter	0.1	0.2	0.3	0.4	0.5	0.6	0.7	0.8	0.9	1.0
Length of Rafter	0.104	0.207	0.311	0.414	0.518	0.621	0.725	0.828	0.932	1.035
Length of Hip										

RISE OF COMMON RAFTER 0.287 m **PER METRE OF RUN**   **PITCH** 16°
(Grecian pitch)

BEVELS: COMMON RAFTER   – SEAT   16
         ″   ″   – RIDGE   74
         HIP OR VALLEY   – SEAT   11.5
         ″   ″   ″ – RIDGE   78.5
         JACK RAFTER   – EDGE   44
         PURLIN   – EDGE   46
         ″   – SIDE   74.5

JACK RAFTERS 333 mm **CENTRES DECREASE** 346 (in mm to 999 and
                 400 ″   ″   ″   416 thereafter in m)
                 500 ″   ″   ″   520
                 600 ″   ″   ″   624

Run of Rafter	0.1	0.2	0.3	0.4	0.5	0.6	0.7	0.8	0.9	1.0
Length of Rafter	0.104	0.208	0.312	0.416	0.52	0.624	0.728	0.832	0.936	1.04
Length of Hip	0.144	0.289	0.433	0.577	0.722	0.866	1.01	1.154	1.299	1.443

RISE OF COMMON RAFTER 0.306 m PER METRE OF RUN   PITCH 17°

BEVELS:
COMMON RAFTER	– SEAT	17	
" "	– RIDGE	73	
HIP OR VALLEY	– SEAT	12	
" " "	– RIDGE	78	
JACK RAFTER	– EDGE	43.5	
PURLIN	– EDGE	46.5	
"	– SIDE	73.5	

JACK RAFTERS 333 mm **CENTRES DECREASE** 348 (in mm to 999 and

400 "	"	"	418	thereafter in m)
500 "	"	"	522	
600 "	"	"	627	

	0.1	0.2	0.3	0.4	0.5	0.6	0.7	0.8	0.9	1.0
Run of Rafter	0.1	0.2	0.3	0.4	0.5	0.6	0.7	0.8	0.9	1.0
Length of Rafter	0.105	0.209	0.314	0.418	0.523	0.627	0.732	0.837	0.941	1.046
Length of Hip	0.145	0.289	0.434	0.579	0.723	0.868	1.013	1.158	1.302	1.447

RISE OF COMMON RAFTER 0.315 m **PER METRE OF RUN**   **PITCH** 17½°

BEVELS:	COMMON RAFTER	– SEAT	17.5
	" "	– RIDGE	72.5
	HIP OR VALLEY	– SEAT	12.5
	" " "	– RIDGE	77.5
	JACK RAFTER	– EDGE	43.5
	PURLIN	– EDGE	46.5
	"	– SIDE	73.5

JACK RAFTERS	333 mm	**CENTRES DECREASE**	349	(in mm to 999 and
	400 "	" "	420	thereafter in m)
	500 "	" "	524	
	600 "	" "	629	

Run of Rafter	0.1	0.2	0.3	0.4	0.5	0.6	0.7	0.8	0.9	1.0
Length of Rafter	0.105	0.210	0.315	0.419	0.524	0.629	0.734	0.839	0.944	1.049
Length of Hip	0.145	0.29	0.435	0.579	0.724	0.829	1.014	1.159	1.304	1.449

RISE OF COMMON RAFTER 0.325 m PER METRE OF RUN   PITCH 18°

BEVELS:　COMMON RAFTER　– SEAT　18
　　　　　　　　　　 "　　　　　 "　　– RIDGE　72
　　　　　　 HIP OR VALLEY　　– SEAT　13
　　　　　　　 "　　 "　　　 "　　 – RIDGE　77
　　　　　　 JACK RAFTER　　 – EDGE　43.5
　　　　　　 PURLIN　　　　　 – EDGE　46.5
　　　　　　　 "　　　　　　　 – SIDE　73

JACK RAFTERS 333 mm **CENTRES DECREASE** 350　(in mm to 999 and
　　　　　　　　　 400　"　　　　"　　　　　　"　　　　421　thereafter in m)
　　　　　　　　　 500　"　　　　"　　　　　　"　　　　526
　　　　　　　　　 600　"　　　　"　　　　　　"　　　　631

Run of Rafter	0.1	0.2	0.3	0.4	0.5	0.6	0.7	0.8	0.9	1.0
Length of Rafter	0.105	0.21	0.315	0.421	0.526	0.631	0.736	0.841	0.946	1.051
Length of Hip	0.145	0.29	0.435	0.58	0.726	0.871	1.016	1.161	1.306	1.451

RISE OF COMMON RAFTER 0.344 m **PER METRE OF RUN**   **PITCH** 19°

BEVELS:	COMMON RAFTER	– SEAT	19
	" "	– RIDGE	71
	HIP OR VALLEY	– SEAT	13.5
	" " "	– RIDGE	76.5
	JACK RAFTER	– EDGE	43.5
	PURLIN	– EDGE	46.5
	"	– SIDE	72

JACK RAFTERS 333 mm **CENTRES DECREASE** 352   (in mm to 999 and
                    400 "   "   "   423   thereafter in m)
                    500 "   "   "   529
                    600 "   "   "   635

Run of Rafter	0.1	0.2	0.3	0.4	0.5	0.6	0.7	0.8	0.9	1.0
Length of Rafter	0.106	0.212	0.317	0.423	0.529	0.635	0.74	0.846	0.952	1.058
Length of Hip	0.146	0.291	0.437	0.582	0.728	0.873	1.019	1.164	1.31	1.456

RISE OF COMMON RAFTER   0.364 m **PER METRE OF RUN**   **PITCH**   20°

BEVELS: COMMON RAFTER   – SEAT   20
           ″   ″   – RIDGE   70
           HIP OR VALLEY   – SEAT   14.5
           ″   ″   ″   – RIDGE   75.5
           JACK RAFTER   – EDGE   43
           PURLIN   – EDGE   47
           ″   – SIDE   71

JACK RAFTERS   333 mm **CENTRES DECREASE**   354   (in mm to 999 and
                   400   ″   ″   ″   426   thereafter in m)
                   500   ″   ″   ″   532
                   600   ″   ″   ″   638

Run of Rafter	0.1	0.2	0.3	0.4	0.5	0.6	0.7	0.8	0.9	1.0
Length of Rafter	0.106	0.213	0.319	0.426	0.532	0.639	0.745	0.851	0.958	1.064
Length of Hip	0.146	0.292	0.438	0.584	0.73	0.876	1.022	1.168	1.314	1.46

RISE OF COMMON RAFTER 0.384 m PER METRE OF RUN   PITCH 21°

BEVELS: COMMON RAFTER   – SEAT   21
                  ″   ″   – RIDGE 69
            HIP OR VALLEY   – SEAT   15
            ″   ″   ″   – RIDGE 75
            JACK RAFTER   – EDGE   43
            PURLIN   – EDGE   47
            ″   – SIDE   70.5

JACK RAFTERS 333 mm **CENTRES DECREASE** 357 (in mm to 999 and
            400 ″   ″   ″   428 thereafter in m)
            500 ″   ″   ″   536
            600 ″   ″   ″   643

Run of Rafter	0.1	0.2	0.3	0.4	0.5	0.6	0.7	0.8	0.9	1.0
Length of Rafter	0.107	0.214	0.321	0.428	0.536	0.643	0.75	0.857	0.964	1.071
Length of Hip	0.147	0.293	0.44	0.586	0.733	0.879	1.026	1.172	1.319	1.465

RISE OF COMMON RAFTER 0.404 m **PER METRE OF RUN**   **PITCH** 22°

	BEVELS:	COMMON RAFTER	– SEAT	22
		″ ″	– RIDGE	68
		HIP OR VALLEY	– SEAT	16
		″ ″ ″	– RIDGE	74
		JACK RAFTER	– EDGE	43
		PURLIN	– EDGE	47
		″	– SIDE	69.5

JACK RAFTERS 333 mm **CENTRES DECREASE** 359   (in mm to 999 and
           400 ″   ″   ″   432   thereafter in m)
           500 ″   ″   ″   540
           600 ″   ″   ″   647

Run of Rafter	0.1	0.2	0.3	0.4	0.5	0.6	0.7	0.8	0.9	1.0
Length of Rafter	0.108	0.216	0.324	0.431	0.539	0.647	0.755	0.863	0.971	1.079
Length of Hip	0.147	0.294	0.441	0.588	0.736	0.883	1.03	1.177	1.324	1.471

RISE OF COMMON RAFTER 0.414 m **PER METRE OF RUN**   **PITCH** $22\frac{1}{2}°$

BEVELS:	COMMON RAFTER	– SEAT	22.5
	" "	– RIDGE	67.5
	HIP OR VALLEY	– SEAT	16.25
	" " "	– RIDGE	73.75
	JACK RAFTER	– EDGE	42.75
	PURLIN	– EDGE	47.5
	"	– SIDE	69.0

JACK RAFTERS 333 mm **CENTRES DECREASE** 361   (in mm to 999 and
          400 "   "   "   433   thereafter in m)
          500 "   "   "   542
          600 "   "   "   650

Run of Rafter	0.1	0.2	0.3	0.4	0.5	0.6	0.7	0.8	0.9	1.0
Length of Rafter	0.108	0.216	0.325	0.433	0.541	0.649	0.758	0.866	0.974	1.082
Length of Hip	0.147	0.294	0.442	0.589	0.737	0.884	1.032	1.179	1.326	1.473

RISE OF COMMON RAFTER 0.424 m **PER METRE OF RUN**   **PITCH** 23°

BEVELS:
COMMON RAFTER	– SEAT	23	
" "	– RIDGE	67	
HIP OR VALLEY	– SEAT	16.5	
" " "	– RIDGE	73.5	
JACK RAFTER	– EDGE	42.5	
PURLIN	– EDGE	47.5	
"	– SIDE	68.5	

JACK RAFTERS   333 mm **CENTRES DECREASE**   362   (in mm to 999 and
     400   "   "   "   434   thereafter in m)
     500   "   "   "   543
     600   "   "   "   652

Run of Rafter	0.1	0.2	0.3	0.4	0.5	0.6	0.7	0.8	0.9	1.0
Length of Rafter	0.109	0.217	0.326	0.435	0.543	0.652	0.76	0.869	0.978	1.086
Length of Hip	0.148	0.295	0.443	0.591	0.739	0.886	1.034	1.181	1.329	1.477

RISE OF COMMON RAFTER 0.445 m **PER METRE OF RUN**   **PITCH** 24°
(Roman pitch)

BEVELS: COMMON RAFTER   – SEAT   24
         "   "   – RIDGE 66
         HIP OR VALLEY   – SEAT   17.5
         "   "   "   – RIDGE 72.5
         JACK RAFTER   – EDGE   42.5
         PURLIN   – EDGE   47.5
         "   – SIDE   68

JACK RAFTERS 333 mm **CENTRES DECREASE** 365   (in mm to 999 and
                  400 "   "   "   438   thereafter in m)
                  500 "   "   "   548
                  600 "   "   "   657

Run of Rafter	0.1	0.2	0.3	0.4	0.5	0.6	0.7	0.8	0.9	1.0
Length of Rafter	0.109	0.219	0.328	0.438	0.547	0.657	0.766	0.876	0.985	1.095
Length of Hip	0.148	0.297	0.445	0.593	0.741	0.89	1.038	1.186	1.334	1.483

RISE OF COMMON RAFTER 0.466 m **PER METRE OF RUN**   **PITCH** 25°

BEVELS:	COMMON RAFTER	– SEAT	25	
	" "	– RIDGE	65	
	HIP OR VALLEY	– SEAT	18	
	" " "	– RIDGE	72	
	JACK RAFTER	– EDGE	42	
	PURLIN	– EDGE	48	
	"	– SIDE	67	

JACK RAFTERS 333 mm **CENTRES DECREASE** 367 (in mm to 999 and
               400 "   "   "   441   thereafter in m)
               500 "   "   "   552
               600 "   "   "   662

Run of Rafter	0.1	0.2	0.3	0.4	0.5	0.6	0.7	0.8	0.9	1.0
Length of Rafter	0.11	0.221	0.331	0.441	0.552	0.662	0.772	0.883	0.993	1.103
Length of Hip	0.149	0.298	0.447	0.596	0.745	0.893	1.042	1.191	1.34	1.489

RISE OF COMMON RAFTER 0.5 m **PER METRE OF RUN**   **PITCH** 26° 34′
                                                             (Quarter pitch)

BEVELS:	COMMON RAFTER	– SEAT	26.5
	″ ″	– RIDGE	63.5
	HIP OR VALLEY	– SEAT	19.5
	″ ″ ″	– RIDGE	70.5
	JACK RAFTER	– EDGE	42
	PURLIN	– EDGE	48
	″	– SIDE	66

JACK RAFTERS 333 mm **CENTRES DECREASE** 372 (in mm to 999 and
                 400 ″   ″   ″   447   thereafter in m)
                 500 ″   ″   ″   559
                 600 ″   ″   ″   671

Run of Rafter	0.1	0.2	0.3	0.4	0.5	0.6	0.7	0.8	0.9	1.0
Length of Rafter	0.112	0.224	0.335	0.447	0.559	0.671	0.783	0.894	1.006	1.118
Length of Hip	0.15	0.3	0.45	0.6	0.75	0.9	1.05	1.2	1.35	1.5

RISE OF COMMON RAFTER 0.521 m **PER METRE OF RUN**   **PITCH** $27\frac{1}{2}°$

BEVELS: COMMON RAFTER   – SEAT   27.5
                 "   "   – RIDGE   62.5
            HIP OR VALLEY   – SEAT   20
            "   "   "   – RIDGE   70
            JACK RAFTER   – EDGE   41.75
            PURLIN   – EDGE   48.5
                 "   – SIDE   65

JACK RAFTERS 333 mm **CENTRES DECREASE** 375   (in mm to 999 and
              400 "   "   "   451   thereafter in m)
              500 "   "   "   563
              600 "   "   "   677

Run of Rafter	0.1	0.2	0.3	0.4	0.5	0.6	0.7	0.8	0.9	1.0
Length of Rafter	0.113	0.225	0.338	0.451	0.564	0.676	0.789	0.902	1.015	1.127
Length of Hip	0.151	0.301	0.452	0.603	0.754	0.904	1.054	1.206	1.365	1.507

RISE OF COMMON RAFTER 0.532 m **PER METRE OF RUN**   **PITCH** 28°

BEVELS:	COMMON RAFTER	– SEAT	28
	" "	– RIDGE	62
	HIP OR VALLEY	– SEAT	20.5
	" " "	– RIDGE	69.5
	JACK RAFTER	– EDGE	41.5
	PURLIN	– EDGE	48.5
	"	– SIDE	65

JACK RAFTERS 333 mm **CENTRES DECREASE** 377   (in mm to 999 and

	400 "	"	"	453	thereafter in m)
	500 "	"	"	566	
	600 "	"	"	680	

Run of Rafter	0.1	0.2	0.3	0.4	0.5	0.6	0.7	0.8	0.9	1.0
Length of Rafter	0.113	0.227	0.34	0.453	0.566	0.68	0.793	0.906	1.019	1.133
Length of Hip	0.151	0.302	0.453	0.603	0.754	0.905	1.056	1.207	1.358	1.511

RISE OF COMMON RAFTER 0.544 m **PER METRE OF RUN**   **PITCH** 29°

BEVELS:
COMMON RAFTER	– SEAT	29	
" "	– RIDGE	61	
HIP OR VALLEY	– SEAT	21.5	
" " "	– RIDGE	68.5	
JACK RAFTER	– EDGE	41	
PURLIN	– EDGE	49	
"	– SIDE	64	

JACK RAFTERS 333 mm **CENTRES DECREASE** 381   (in mm to 999 and

400 "	"	"	457	thereafter in m)
500 "	"	"	572	
600 "	"	"	686	

Run of Rafter	0.1	0.2	0.3	0.4	0.5	0.6	0.7	0.8	0.9	1.0
Length of Rafter	0.114	0.229	0.343	0.457	0.572	0.686	0.8	0.914	1.029	1.143
Length of Hip	0.152	0.304	0.456	0.608	0.759	0.912	1.063	1.215	1.367	1.519

RISE OF COMMON RAFTER 0.577 m **PER METRE OF RUN**   **PITCH** 30°

BEVELS:	COMMON RAFTER	– SEAT	30
	" "	– RIDGE	60
	HIP OR VALLEY	– SEAT	22
	" " "	– RIDGE	68
	JACK RAFTER	– EDGE	41
	PURLIN	– EDGE	49
	"	– SIDE	63.5

JACK RAFTERS 333 mm **CENTRES DECREASE** 385 (in mm to 999 and
            400 "   "   "   462   thereafter in m)
            500 "   "   "   577
            600 "   "   "   693

Run of Rafter	0.1	0.2	0.3	0.4	0.5	0.6	0.7	0.8	0.9	1.0
Length of Rafter	0.116	0.231	0.346	0.462	0.577	0.693	0.808	0.924	1.039	1.155
Length of Hip	0.153	0.306	0.458	0.611	0.764	0.917	1.069	1.222	1.375	1.528

RISE OF COMMON RAFTER 0.601 m **PER METRE OF RUN**   **PITCH** 31°

BEVELS: COMMON RAFTER – SEAT 31
                "   "   – RIDGE 59
             HIP OR VALLEY – SEAT 23
           "   "   "   – RIDGE 67
             JACK RAFTER – EDGE 40.5
             PURLIN – EDGE 49.5
           "   – SIDE 62.5

JACK RAFTERS 333 mm **CENTRES DECREASE** 389 (in mm to 999 and
           400 "   "   "   467   thereafter in m)
           500 "   "   "   584
           600 "   "   "   700

Run of Rafter	0.1	0.2	0.3	0.4	0.5	0.6	0.7	0.8	0.9	1.0
Length of Rafter	0.117	0.233	0.35	0.467	0.583	0.7	0.817	0.933	1.05	1.167
Length of Hip	0.154	0.307	0.461	0.615	0.768	0.922	1.076	1.229	1.383	1.537

RISE OF COMMON RAFTER 0.625 m **PER METRE OF RUN**   **PITCH** 32°

BEVELS: COMMON RAFTER   – SEAT 32
                           ″   ″   – RIDGE 58
                  HIP OR VALLEY   – SEAT 24
                  ″   ″   ″   – RIDGE 66
                  JACK RAFTER   – EDGE 40.5
                  PURLIN   – EDGE 49.5
                  ″   – SIDE 62

JACK RAFTERS 333 mm **CENTRES DECREASE** 393   (in mm to 999 and
                    400 ″   ″   ″   472   thereafter in m)
                    500 ″   ″   ″   590
                    600 ″   ″   ″   707

Run of Rafter	0.1	0.2	0.3	0.4	0.5	0.6	0.7	0.8	0.9	1.0
Length of Rafter	0.118	0.239	0.354	0.472	0.59	0.708	0.825	0.943	1.061	1.179
Length of Hip	0.155	0.309	0.464	0.618	0.773	0.928	1.082	1.237	1.391	1.546

RISE OF COMMON RAFTER 0.637 m **PER METRE OF RUN**   **PITCH** $32\frac{1}{2}°$

BEVELS: COMMON RAFTER   – SEAT 32.5
                "   "   – RIDGE 57.5
         HIP OR VALLEY   – SEAT 24.25
         "   "   "   – RIDGE 65.75
         JACK RAFTER   – EDGE 40.25
         PURLIN   – EDGE 49.75
         "   – SIDE 61.5

JACK RAFTERS 333 mm **CENTRES DECREASE** 395 (in mm to 999 and
                 400 "   "   "   475 thereafter in m)
                 500 "   "   "   593
                 600 "   "   "   711

Run of Rafter	0.1	0.2	0.3	0.4	0.5	0.6	0.7	0.8	0.9	1.0
Length of Rafter	0.119	0.237	0.356	0.474	0.593	0.711	0.830	0.949	1.067	1.186
Length of Hip	0.155	0.310	0.466	0.620	0.776	0.930	1.086	1.241	1.396	1.551

RISE OF COMMON RAFTER 0.649 m **PER METRE OF RUN**   **PITCH** 33°

BEVELS:	COMMON RAFTER	– SEAT	33
	" "	– RIDGE	57
	HIP OR VALLEY	– SEAT	24.5
	" " "	– RIDGE	65.5
	JACK RAFTER	– EDGE	40
	PURLIN	– EDGE	50
	"	– SIDE	61.5

JACK RAFTERS 333 mm **CENTRES DECREASE** 397 (in mm to 999 and
          400 "   "   "   477   thereafter in m)
          500 "   "   "   596
          600 "   "   "   715

Run of Rafter	0.1	0.2	0.3	0.4	0.5	0.6	0.7	0.8	0.9	1.0
Length of Rafter	0.119	0.238	0.358	0.48	0.596	0.715	0.835	0.954	1.073	1.192
Length of Hip	0.156	0.311	0.467	0.623	0.778	0.934	1.089	1.245	1.401	1.556

RISE OF COMMON RAFTER 0.666 m PER METRE OF RUN

PITCH 33° 40′
(Third pitch)

BEVELS: COMMON RAFTER   – SEAT 33.5
                   ″   ″   – RIDGE 56.5
           HIP OR VALLEY   – SEAT 25
              ″   ″   ″   – RIDGE 65
           JACK RAFTER   – EDGE 40
           PURLIN   – EDGE 50
              ″   – SIDE 61

JACK RAFTERS 333 mm **CENTRES DECREASE** 397 (in mm to 999 and
                   400 ″   ″   ″   481 thereafter in m)
                   500 ″   ″   ″   601
                   600 ″   ″   ″   721

Run of Rafter	0.1	0.2	0.3	0.4	0.5	0.6	0.7	0.8	0.9	1.0
Length of Rafter	0.12	0.24	0.361	0.481	0.601	0.721	0.841	0.961	1.082	1.202
Length of Hip	0.157	0.313	0.47	0.626	0.782	0.938	1.094	1.251	1.408	1.563

RISE OF COMMON RAFTER 0.7 m PER METRE OF RUN   PITCH 35°

BEVELS: COMMON RAFTER   – SEAT 35
                                "   "   – RIDGE 55
                    HIP OR VALLEY   – SEAT 26.5
                    "   "   "   – RIDGE 63.5
                    JACK RAFTER   – EDGE 39.5
                    PURLIN   – EDGE 50.5
                    "   – SIDE 60

JACK RAFTERS 333 mm **CENTRES DECREASE** 407 (in mm to 999 and thereafter in m)

333 mm	"	"	407
400 "	"	"	488
500 "	"	"	611
600 "	"	"	733

Run of Rafter	0.1	0.2	0.3	0.4	0.5	0.6	0.7	0.8	0.9	1.0
Length of Rafter	0.122	0.244	0.366	0.488	0.61	0.732	0.855	0.977	1.099	1.221
Length of Hip	0.158	0.316	0.473	0.631	0.789	0.947	1.105	1.262	1.42	1.578

RISE OF COMMON RAFTER 0.727 m PER METRE OF RUN   PITCH 36°

BEVELS: COMMON RAFTER   – SEAT   36
               "   "   – RIDGE 54
           HIP OR VALLEY   – SEAT   27
             "   "   "   – RIDGE 63
           JACK RAFTER   – EDGE   39
           PURLIN   – EDGE   51
               "   – SIDE   59.5

JACK RAFTERS 333 mm **CENTRES DECREASE** 412   (in mm to 999 and
                  400 "   "   "   494   thereafter in m)
                  500 "   "   "   618
                  600 "   "   "   742

Run of Rafter	0.1	0.2	0.3	0.4	0.5	0.6	0.7	0.8	0.9	1.0
Length of Rafter	0.124	0.247	0.371	0.494	0.618	0.742	0.865	0.989	1.112	1.236
Length of Hip	0.159	0.318	0.477	0.636	0.795	0.954	1.113	1.272	1.431	1.59

RISE OF COMMON RAFTER 0.754 m **PER METRE OF RUN**   **PITCH** 37°

BEVELS: COMMON RAFTER   – SEAT   37
              ″   ″   – RIDGE 53
         HIP OR VALLEY   – SEAT   28
           ″   ″   ″   – RIDGE 62
         JACK RAFTER   – EDGE   38.5
         PURLIN   – EDGE   51.5
              ″   – SIDE   59

JACK RAFTERS 333 mm **CENTRES DECREASE** 417   (in mm to 999 and
                  400 ″   ″   ″   501   thereafter in m)
                  500 ″   ″   ″   626
                  600 ″   ″   ″   751

	Run of Rafter	0.1	0.2	0.3	0.4	0.5	0.6	0.7	0.8	0.9	1.0
Length of Rafter	0.125	0.25	0.376	0.501	0.626	0.751	0.876	1.002	1.127	1.252
Length of Hip	0.16	0.32	0.481	0.641	0.801	0.961	1.122	1.282	1.442	1.602

RISE OF COMMON RAFTER 0.767 m **PER METRE OF RUN**   **PITCH** 37½°

BEVELS: COMMON RAFTER   – SEAT   37.5
            "   "   – RIDGE 52.5
         HIP OR VALLEY   – SEAT   28.5
            "   "   "   – RIDGE 61.5
         JACK RAFTER   – EDGE   38.25
         PURLIN   – EDGE   51.75
            "   – SIDE   58.5

JACK RAFTERS 333 mm **CENTRES DECREASE** 420   (in mm to 999 and
                  400 "   "   "   505   thereafter in m)
                  500 "   "   "   630
                  600 "   "   "   757

Run of Rafter	0.1	0.2	0.3	0.4	0.5	0.6	0.7	0.8	0.9	1.0
Length of Rafter	0.126	0.252	0.378	0.504	0.630	0.756	0.882	1.008	1.134	1.260
Length of Hip	0.161	0.322	0.482	0.636	0.804	1.965	1.126	1.286	1.448	1.609

RISE OF COMMON RAFTER 0.781 m **PER METRE OF RUN**   PITCH 38°

BEVELS: COMMON RAFTER   – SEAT   38
                ″   ″   – RIDGE 52
          HIP OR VALLEY   – SEAT   29
            ″   ″   ″   – RIDGE 61
          JACK RAFTER   – EDGE   38
          PURLIN   – EDGE   52
               ″   – SIDE   58.5

JACK RAFTERS 333 mm **CENTRES DECREASE** 423   (in mm to 999 and
                  400 ″   ″   ″   508   thereafter in m)
                  500 ″   ″   ″   635
                  600 ″   ″   ″   761

Run of Rafter	0.1	0.2	0.3	0.4	0.5	0.6	0.7	0.8	0.9	1.0
Length of Rafter	0.127	0.254	0.381	0.508	0.635	0.761	0.888	1.015	1.142	1.269
Length of Hip	0.162	0.323	0.485	0.646	0.808	0.969	1.131	1.293	1.454	1.616

RISE OF COMMON RAFTER 0.81m **PER METRE OF RUN**   **PITCH** 39°

BEVELS: COMMON RAFTER   – SEAT   39
              "   "   – RIDGE   51
            HIP OR VALLEY   – SEAT   30
           "   "   "   – RIDGE   60
            JACK RAFTER   – EDGE   38
            PURLIN   – EDGE   52
           "   – SIDE   58

JACK RAFTERS 333 mm **CENTRES DECREASE** 429   (in mm to 999 and
               400 "   "   "   515   thereafter in m)
               500 "   "   "   644
               600 "   "   "   772

Run of Rafter	0.1	0.2	0.3	0.4	0.5	0.6	0.7	0.8	0.9	1.0
Length of Rafter	0.129	0.257	0.386	0.515	0.643	0.772	0.901	1.029	1.158	1.287
Length of Hip	0.163	0.326	0.489	0.652	0.815	0.978	1.141	1.304	1.467	1.63

RISE OF COMMON RAFTER 0.839 m **PER METRE OF RUN**   **PITCH** 40°

BEVELS: COMMON RAFTER   – SEAT 40
             ″   ″   – RIDGE 50
        HIP OR VALLEY   – SEAT 30.5
        ″   ″   ″   – RIDGE 59.5
        JACK RAFTER   – EDGE 37.5
        PURLIN   – EDGE 52.5
           ″   – SIDE 57.5

JACK RAFTERS   333 mm **CENTRES DECREASE**   435   (in mm to 999 and
              400 ″   ″   ″   522   thereafter in m)
              500 ″   ″   ″   653
              600 ″   ″   ″   783

Run of Rafter	0.1	0.2	0.3	0.4	0.5	0.6	0.7	0.8	0.9	1.0
Length of Rafter	0.131	0.261	0.392	0.522	0.653	0.783	0.914	1.044	1.175	1.305
Length of Hip	0.164	0.329	0.493	0.658	0.822	0.987	1.151	1.316	1.48	1.644

RISE OF COMMON RAFTER 0.869 m **PER METRE OF RUN**   **PITCH** 41°

BEVELS: COMMON RAFTER   – SEAT   41
                                     ″   ″   – RIDGE 49
                HIP OR VALLEY   – SEAT   31.5
                 ″   ″   ″   – RIDGE 58.5
                JACK RAFTER   – EDGE 37
                PURLIN   – EDGE 53
                    ″   – SIDE   56.5

JACK RAFTERS 333 mm **CENTRES DECREASE** 441 (in mm to 999 and
              400 ″   ″   ″   530   thereafter in m)
              500 ″   ″   ″   663
              600 ″   ″   ″   795

Run of Rafter	0.1	0.2	0.3	0.4	0.5	0.6	0.7	0.8	0.9	1.0
Length of Rafter	0.133	0.265	0.398	0.53	0.663	0.795	0.928	1.06	1.193	1.325
Length of Hip	0.166	0.332	0.498	0.664	0.83	0.996	1.162	1.328	1.494	1.66

RISE OF COMMON RAFTER 0.9 m **PER METRE OF RUN**   **PITCH** 42°

BEVELS: COMMON RAFTER   – SEAT 42
                    ″   ″   – RIDGE 48
                HIP OR VALLEY   – SEAT 32.5
                ″   ″   ″   – RIDGE 57.5
                JACK RAFTER   – EDGE 36.5
                PURLIN   – EDGE 53.5
                    ″   – SIDE 56

JACK RAFTERS 333 mm **CENTRES DECREASE** 448 (in mm to 999 and
               400 ″   ″   ″   538   thereafter in m)
               500 ″   ″   ″   673
               600 ″   ″   ″   808

Run of Rafter	0.1	0.2	0.3	0.4	0.5	0.6	0.7	0.8	0.9	1.0
Length of Rafter	0.135	0.269	0.404	0.538	0.673	0.807	0.942	1.097	1.211	1.346
Length of Hip	0.168	0.335	0.503	0.671	0.838	1.006	1.173	1.341	1.509	1.677

RISE OF COMMON RAFTER 0.916 m **PER METRE OF RUN**   **PITCH** 42½°

BEVELS: COMMON RAFTER   – SEAT 42.5
                   "   "   – RIDGE 47.5
          HIP OR VALLEY   – SEAT 33
             "   "   "   – RIDGE 57
          JACK RAFTER   – EDGE 36.25
          PURLIN   – EDGE 53.75
               "   – SIDE 55.75

JACK RAFTERS 333 mm **CENTRES DECREASE** 452 (in mm to 999 and
          400 "   "   "   543 thereafter in m)
          500 "   "   "   679
          600 "   "   "   815

Run of Rafter	0.1	0.2	0.3	0.4	0.5	0.6	0.7	0.8	0.9	1.0
Length of Rafter	0.137	0.271	0.406	0.543	0.678	0.814	0.949	1.085	1.221	1.356
Length of Hip	0.170	0.337	0.505	0.674	0.842	1.011	1.179	1.348	1.569	1.685

RISE OF COMMON RAFTER 0.933 m **PER METRE OF RUN**   **PITCH** 43°

BEVELS:	COMMON RAFTER	– SEAT	43
	" "	– RIDGE	47
	HIP OR VALLEY	– SEAT	33.5
	" " "	– RIDGE	56.5
	JACK RAFTER	– EDGE	36
	PURLIN	– EDGE	54
	"	– SIDE	55.5

JACK RAFTERS 333 mm **CENTRES DECREASE** 455   (in mm to 999 and
          400 "   "   "   547   thereafter in m)
          500 "   "   "   684
          600 "   "   "   820

Run of Rafter	0.1	0.2	0.3	0.4	0.5	0.6	0.7	0.8	0.9	1.0
Length of Rafter	0.137	0.273	0.41	0.547	0.684	0.82	0.967	1.094	1.231	1.367
Length of Hip	0.169	0.339	0.508	0.678	0.847	1.016	1.186	1.355	1.525	1.694

RISE OF COMMON RAFTER 0.966 m **PER METRE OF RUN**   **PITCH** 44°

BEVELS: COMMON RAFTER — SEAT 44
               ″   ″ — RIDGE 46
            HIP OR VALLEY — SEAT 34.5
           ″   ″   ″ — RIDGE 55.5
            JACK RAFTER — EDGE 35.5
            PURLIN — EDGE 54.5
           ″   — SIDE 55

JACK RAFTERS 333 mm **CENTRES DECREASE** 463 (in mm to 999 and
             400 ″   ″   ″   556 thereafter in m)
             500 ″   ″   ″   695
             600 ″   ″   ″   834

Run of Rafter	0.1	0.2	0.3	0.4	0.5	0.6	0.7	0.8	0.9	1.0
Length of Rafter	0.139	0.278	0.417	0.556	0.695	0.834	0.973	1.111	1.251	1.39
Length of Hip	0.171	0.342	0.514	0.685	0.856	1.027	1.199	1.37	1.541	1.712

RISE OF COMMON RAFTER 1.0 m **PER METRE OF RUN**   **PITCH** 45°

	BEVELS:	COMMON RAFTER	– SEAT	45
		" "	– RIDGE	45
		HIP OR VALLEY	– SEAT	35.5
		" " "	– RIDGE	54.5
		JACK RAFTER	– EDGE	35.5
		PURLIN	– EDGE	54.5
		"	– SIDE	54.5

JACK RAFTERS 333 mm **CENTRES DECREASE** 471   (in mm to 999 and
400 "   "   "   566   thereafter in m)
500 "   "   "   707
600 "   "   "   848

Run of Rafter	0.1	0.2	0.3	0.4	0.5	0.6	0.7	0.8	0.9	1.0
Length of Rafter	0.141	0.283	0.424	0.566	0.707	0.848	0.99	1.131	1.273	1.414
Length of Hip	0.173	0.346	0.519	0.693	0.866	1.039	1.212	1.386	1.559	1.732

RISE OF COMMON RAFTER 1.036 m **PER METRE OF RUN**   **PITCH** 46°

BEVELS: COMMON RAFTER   – SEAT   46
                   ″   ″   – RIDGE 44
         HIP OR VALLEY   – SEAT   36
           ″   ″   ″   – RIDGE 54
         JACK RAFTER   – EDGE   35
         PURLIN   – EDGE   55
            ″   – SIDE   54.5

JACK RAFTERS 333 mm **CENTRES DECREASE** 480   (in mm to 999 and
                 400 ″   ″   ″   576   thereafter in m)
                 500 ″   ″   ″   720
                 600 ″   ″   ″   864

Run of Rafter	0.1	0.2	0.3	0.4	0.5	0.6	0.7	0.8	0.9	1.0
Length of Rafter	0.144	0.288	0.432	0.576	0.72	0.864	1.058	1.152	1.296	1.44
Length of Hip	0.175	0.351	0.526	0.701	0.876	1.052	1.227	1.402	1.578	1.753

RISE OF COMMON RAFTER 1.072 m PER METRE OF RUN   PITCH 47°

BEVELS: COMMON RAFTER   – SEAT   47
       //   //   – RIDGE   43
    HIP OR VALLEY   – SEAT   37
    //   //   //   – RIDGE   53
    JACK RAFTER   – EDGE   34.5
    PURLIN   – EDGE   55.5
    //   – SIDE   54

JACK RAFTERS   333 mm **CENTRES DECREASE**   488   (in mm to 999 and
       400   //   //   //   586   thereafter in m)
       500   //   //   //   733
       600   //   //   //   880

Run of Rafter	0.1	0.2	0.3	0.4	0.5	0.6	0.7	0.8	0.9	1.0
Length of Rafter	0.147	0.293	0.44	0.587	0.733	0.88	1.026	1.173	1.32	1.466
Length of Hip	0.177	0.355	0.532	0.71	0.887	1.065	1.242	1.42	1.597	1.775

RISE OF COMMON RAFTER 1.111 m **PER METRE OF RUN**   **PITCH** 48°

BEVELS: COMMON RAFTER   – SEAT   48
                "   "   – RIDGE 42
              HIP OR VALLEY   – SEAT   38
             "   "   "   – RIDGE 52
              JACK RAFTER   – EDGE 34
              PURLIN   – EDGE 56
                "   – SIDE   53.5

JACK RAFTERS 333 mm **CENTRES DECREASE** 498   (in mm to 999 and
               400 "   "   "   598   thereafter in m)
               500 "   "   "   747
               600 "   "   "   896

Run of Rafter	0.1	0.2	0.3	0.4	0.5	0.6	0.7	0.8	0.9	1.0
Length of Rafter	0.149	0.299	0.448	0.598	0.747	0.897	1.046	1.196	1.345	1.494
Length of Hip	0.18	0.36	0.539	0.719	0.899	1.079	1.259	1.438	1.618	1.798

RISE OF COMMON RAFTER 1.15 m **PER METRE OF RUN**   **PITCH** 49°

BEVELS: COMMON RAFTER   – SEAT   49
                                           ''   ''   – RIDGE 41
                HIP OR VALLEY   – SEAT   39
                ''   ''   ''   – RIDGE 51
                JACK RAFTER   – EDGE   33.5
                PURLIN   – EDGE   56.5
                ''   – SIDE   53

JACK RAFTERS 333 mm **CENTRES DECREASE** 508 (in mm to 999 and
                    400 ''   ''   ''   610 thereafter in m)
                    500 ''   ''   ''   762
                    600 ''   ''   ''   914

Run of Rafter	0.1	0.2	0.3	0.4	0.5	0.6	0.7	0.8	0.9	1.0
Length of Rafter	0.152	0.305	0.457	0.61	0.762	0.915	1.067	1.219	1.372	1.524
Length of Hip	0.182	0.365	0.547	0.729	0.912	1.094	1.276	1.458	1.641	1.823

RISE OF COMMON RAFTER 1.192 m **PER METRE OF RUN**   **PITCH** 50°

BEVELS: COMMON RAFTER   – SEAT 50
                 ″   ″   – RIDGE 40
               HIP OR VALLEY   – SEAT   40
              ″   ″   ″   – RIDGE 50
              JACK RAFTER   – EDGE   32.5
              PURLIN   – EDGE   57.5
              ″   – SIDE   52.5

JACK RAFTERS 333 mm **CENTRES DECREASE** 518 (in mm to 999 and
                400 ″   ″   ″   622   thereafter in m)
                500 ″   ″   ″   778
                600 ″   ″   ″   934

Run of Rafter	0.1	0.2	0.3	0.4	0.5	0.6	0.7	0.8	0.9	1.0
Length of Rafter	0.156	0.311	0.467	0.622	0.778	0.933	1.089	1.246	1.4	1.556
Length of Hip	0.185	0.37	0.555	0.74	0.925	1.11	1.295	1.48	1.664	1.849

RISE OF COMMON RAFTER 1.235 m **PER METRE OF RUN**   **PITCH** 51°

BEVELS: COMMON RAFTER   – SEAT   51
          "   "   – RIDGE 39
         HIP OR VALLEY   – SEAT   41
         "   "   "   – RIDGE 49
         JACK RAFTER   – EDGE   32
         PURLIN   – EDGE   58
         "   – SIDE   52

JACK RAFTERS   333 mm **CENTRES DECREASE**   529   (in mm to 999 and
                400   "   "   "   636   thereafter in m)
                500   "   "   "   795
                600   "   "   "   953

Run of Rafter	0.1	0.2	0.3	0.4	0.5	0.6	0.7	0.8	0.9	1.0
Length of Rafter	0.159	0.318	0.477	0.636	0.795	0.953	1.012	1.271	1.43	1.589
Length of Hip	0.188	0.375	0.563	0.751	0.936	1.126	1.314	1.502	1.69	1.877

RISE OF COMMON RAFTER 1.28 m **PER METRE OF RUN**   **PITCH** 52°

BEVELS: COMMON RAFTER   – SEAT   52
          "   "   – RIDGE 38
          HIP OR VALLEY   – SEAT   42
          "   "   "   – RIDGE 48
          JACK RAFTER   – EDGE   31.5
          PURLIN   – EDGE   58.5
          "   – SIDE   52

JACK RAFTERS 333 mm **CENTRES DECREASE** 541   (in mm to 999 and
                  400 "   "   "   650   thereafter in m)
                  500 "   "   "   812
                  600 "   "   "   974

Run of Rafter	0.1	0.2	0.3	0.4	0.5	0.6	0.7	0.8	0.9	1.0
Length of Rafter	0.162	0.325	0.487	0.65	0.812	0.974	1.137	1.299	1.462	1.624
Length of Hip	0.191	0.381	0.572	0.763	0.954	1.144	1.335	1.526	1.717	1.907

RISE OF COMMON RAFTER 1.327 m **PER METRE OF RUN**   **PITCH** 53°

BEVELS: COMMON RAFTER   – SEAT   53
                   ʺ   ʺ   – RIDGE   37
             HIP OR VALLEY   – SEAT   43
             ʺ   ʺ   ʺ   – RIDGE   47
             JACK RAFTER   – EDGE   31
             PURLIN   – EDGE   59
                  ʺ   – SIDE   51.5

JACK RAFTERS 333 mm **CENTRES DECREASE** 553   (in mm to 999 and
                 400   ʺ   ʺ   ʺ   665   thereafter in m)
                 500   ʺ   ʺ   ʺ   831
                 600   ʺ   ʺ   ʺ   997

Run of Rafter	0.1	0.2	0.3	0.4	0.5	0.6	0.7	0.8	0.9	1.0
Length of Rafter	0.166	0.332	0.498	0.665	0.831	0.997	1.163	1.329	1.495	1.662
Length of Hip	0.194	0.388	0.582	0.776	0.97	1.164	1.358	1.551	1.745	1.939

RISE OF COMMON RAFTER 1.376 m **PER METRE OF RUN**   **PITCH** 54°

BEVELS:	COMMON RAFTER	– SEAT	54
	" "	– RIDGE	36
	HIP OR VALLEY	– SEAT	44
	" " "	– RIDGE	46
	JACK RAFTER	– EDGE	30.5
	PURLIN	– EDGE	59.5
	"	– SIDE	51

JACK RAFTERS 333 mm **CENTRES DECREASE** 567 (in mm to 999 and
400 "   "   "   680   thereafter in m)
500 "   "   "   850
600 "   "   "   1.021

Run of Rafter	0.1	0.2	0.3	0.4	0.5	0.6	0.7	0.8	0.9	1.0
Length of Rafter	0.17	0.34	0.51	0.681	0.851	1.021	1.191	1.361	1.531	1.701
Length of Hip	0.197	0.395	0.592	0.789	0.987	1.184	1.381	1.579	1.776	1.973

RISE OF COMMON RAFTER 1.428 m **PER METRE OF RUN**   **PITCH** 55°

BEVELS: COMMON RAFTER   – SEAT   55
                  ″   ″   – RIDGE 35
          HIP OR VALLEY   – SEAT   45.5
          ″   ″   ″   – RIDGE 44.5
          JACK RAFTER   – EDGE   30
          PURLIN   – EDGE   60
                  ″   – SIDE   50.5

JACK RAFTERS 333 mm **CENTRES DECREASE** 580 (in mm to 999 and
                  400 ″   ″   ″   697 thereafter in m)
                  500 ″   ″   ″   872
                  600 ″   ″   ″   1.046

Run of Rafter	0.1	0.2	0.3	0.4	0.5	0.6	0.7	0.8	0.9	1.0
Length of Rafter	0.174	0.349	0.523	0.697	0.872	1.046	1.22	1.395	1.569	1.743
Length of Hip	0.201	0.402	0.603	0.804	1.005	1.206	1.407	1.608	1.809	2.01

RISE OF COMMON RAFTER   1.5 m **PER METRE OF RUN**

PITCH 56° 18′
(Italian pitch)

BEVELS: COMMON RAFTER   – SEAT   56.5
      ″   ″   – RIDGE 33.5
     HIP OR VALLEY   – SEAT   46.5
     ″ ″   ″   – RIDGE 43.5
     JACK RAFTER   – EDGE   29
     PURLIN   – EDGE   61
     ″   – SIDE   50

JACK RAFTERS 333 mm **CENTRES DECREASE**   600   (in mm to 999 and
      400 ″   ″   ″   720   thereafter in m)
      500 ″   ″   ″   900
      600 ″   ″   ″   1.082

Run of Rafter	0.1	0.2	0.3	0.4	0.5	0.6	0.7	0.8	0.9	1.0
Length of Rafter	0.18	0.361	0.541	0.721	0.902	1.082	1.262	1.442	1.623	1.803
Length of Hip	0.206	0.412	0.618	0.824	1.031	1.237	1.443	1.649	1.855	2.061

RISE OF COMMON RAFTER 1.6 m **PER METRE OF RUN**   **PITCH** 58°

BEVELS: COMMON RAFTER   – SEAT 58
              ″   ″   – RIDGE 32
         HIP OR VALLEY   – SEAT 48.5
              ″   ″   ″   – RIDGE 41.5
         JACK RAFTER   – EDGE 28
         PURLIN   – EDGE 62
              ″   – SIDE 49.5

JACK RAFTERS 333 mm **CENTRES DECREASE** 628 (in mm to 999 and
                 400 ″   ″   ″   755 thereafter in m)
                 500 ″   ″   ″   944
                 600 ″   ″   ″   1.132

Run of Rafter	0.1	0.2	0.3	0.4	0.5	0.6	0.7	0.8	0.9	1.0
Length of Rafter	0.189	0.377	0.566	0.755	0.944	1.132	1.321	1.51	1.698	1.887
Length of Hip	0.214	0.428	0.641	0.855	1.069	1.283	1.496	1.71	1.924	2.136

RISE OF COMMON RAFTER 1.664 m **PER METRE OF RUN**   **PITCH** 59°

BEVELS:	COMMON RAFTER	– SEAT	59
	" "	– RIDGE	31
	HIP OR VALLEY	– SEAT	49.5
	" " "	– RIDGE	40.5
	JACK RAFTER	– EDGE	27.5
	PURLIN	– EDGE	62.5
	"	– SIDE	49.5

JACK RAFTERS 333 mm **CENTRES DECREASE**   647   (in mm to 999 and
          400 "   "   "   777   thereafter in m)
          500 "   "   "   971
          600 "   "   "   1.165

Run of Rafter	0.1	0.2	0.3	0.4	0.5	0.6	0.7	0.8	0.9	1.0
Length of Rafter	0.194	0.398	0.582	0.777	0.971	1.165	1.359	1.553	1.747	1.942
Length of Hip	0.218	0.437	0.655	0.874	1.092	1.31	1.529	1.747	1.965	2.184

RISE OF COMMON RAFTER 1.732 m **PER METRE OF RUN**   **PITCH** 60°
(Equilateral pitch)

BEVELS: COMMON RAFTER   – SEAT   60
          "   " – RIDGE   30
          HIP OR VALLEY   – SEAT   51
          "   "   "   – RIDGE   39
          JACK RAFTER   – EDGE   26.5
          PURLIN   – EDGE   63.5
          "   – SIDE   49

JACK RAFTERS 333 mm **CENTRES DECREASE**   666   (in mm to 999 and
                  400 "   "   "   800   thereafter in m)
                  500 "   "   "   1.000
                  600 "   "   "   1.200

Run of Rafter	0.1	0.2	0.3	0.4	0.5	0.6	0.7	0.8	0.9	1.0
Length of Rafter	0.2	0.4	0.6	0.8	1.0	1.2	1.4	1.6	1.8	2.0
Length of Hip	0.224	0.447	0.671	0.894	1.118	1.342	1.565	1.789	2.012	2.236

RISE OF COMMON RAFTER 1.804 m **PER METRE OF RUN**   **PITCH** 61°

BEVELS: COMMON RAFTER   – SEAT   61
                                                           ″   ″   – RIDGE   29
                  HIP OR VALLEY   – SEAT   52
                  ″   ″   ″   – RIDGE   38
                  JACK RAFTER   – EDGE   26
                  PURLIN   – EDGE   64
                         ″   – SIDE   49

JACK RAFTERS 333 mm **CENTRES DECREASE**   687   (in mm to 999 and
                  400 ″   ″   ″   825   thereafter in m)
                  500 ″   ″   ″   1.032
                  600 ″   ″   ″   1.238

Run of Rafter	0.1	0.2	0.3	0.4	0.5	0.6	0.7	0.8	0.9	1.0
Length of Rafter	0.206	0.413	0.619	0.825	1.031	1.238	1.444	1.65	1.857	2.063
Length of Hip	0.229	0.458	0.688	0.917	1.146	1.375	1.605	1.834	2.063	2.292

RISE OF COMMON RAFTER   1.88 m **PER METRE OF RUN**   **PITCH**   62°

BEVELS:	COMMON RAFTER		– SEAT	62
	"	"	– RIDGE	28
	HIP OR VALLEY		– SEAT	53
	"	" "	– RIDGE	37
	JACK RAFTER		– EDGE	25
	PURLIN		– EDGE	65
	"		– SIDE	48.5

JACK RAFTERS   333 mm   **CENTRES DECREASE**   709   (in mm to 999 and

	400 "	"	"	852	thereafter in m)
	500 "	"	"	1.065	
	600 "	"	"	1.278	

	0.1	0.2	0.3	0.4	0.5	0.6	0.7	0.8	0.9	1.0
Run of Rafter	0.1	0.2	0.3	0.4	0.5	0.6	0.7	0.8	0.9	1.0
Length of Rafter	0.213	0.426	0.639	0.852	1.065	1.278	1.491	1.704	1.917	2.13
Length of Hip	0.235	0.471	0.706	0.941	1.177	1.412	1.647	1.882	2.118	2.353

RISE OF COMMON RAFTER 2.0 m **PER METRE OF RUN**   **PITCH** 63° 26′
(Gothic pitch)

BEVELS: COMMON RAFTER – SEAT 63.5
　　　　　　　〃　　　　〃 – RIDGE 26.5
　　　　HIP OR VALLEY – SEAT 54.5
　　　　〃　　〃　　〃 – RIDGE 35.5
　　　　JACK RAFTER – EDGE 24
　　　　PURLIN – EDGE 66
　　　　　　〃 – SIDE 48

JACK RAFTERS 333 mm **CENTRES DECREASE** 745 (in mm to 999 and
　　　　　　　　400　〃　　　〃　　　　〃 894 thereafter in m)
　　　　　　　　500　〃　　　〃　　　　〃 1.118
　　　　　　　　600　〃　　　〃　　　　〃 1.342

Run of Rafter	0.1	0.2	0.3	0.4	0.5	0.6	0.7	0.8	0.9	1.0
Length of Rafter	0.224	0.447	0.671	0.894	1.118	1.342	1.565	1.789	2.012	2.236
Length of Hip	0.245	0.49	0.735	0.98	1.225	1.47	1.715	1.96	2.205	2.45

RISE OF COMMON RAFTER 2.145 m **PER METRE OF RUN**   **PITCH** 65°

BEVELS:	COMMON RAFTER	– SEAT	65
	" "	– RIDGE	25
	HIP OR VALLEY	– SEAT	56.5
	" " "	– RIDGE	33.5
	JACK RAFTER	– EDGE	23
	PURLIN	– EDGE	67
	"	– SIDE	48

JACK RAFTERS 333 mm **CENTRES DECREASE**   788   (in mm to 999 and
                     400 "   "   "   946   thereafter in m)
                     500 "   "   "   1.183
                     600 "   "   "   1.420

Run of Rafter	0.1	0.2	0.3	0.4	0.5	0.6	0.7	0.8	0.9	1.0
Length of Rafter	0.237	0.473	0.71	0.946	1.183	1.42	1.656	1.893	2.13	2.366
Length of Hip	0.257	0.514	0.771	1.028	1.284	1.541	1.798	2.055	2.312	2.569

RISE OF COMMON RAFTER 2.246 m **PER METRE OF RUN**   **PITCH** 66°

BEVELS:	COMMON RAFTER	– SEAT	66
	" "	– RIDGE	24
	HIP OR VALLEY	– SEAT	58
	" " "	– RIDGE	32
	JACK RAFTER	– EDGE	22
	PURLIN	– EDGE	68
	"	– SIDE	47.5

JACK RAFTERS 333 mm **CENTRES DECREASE**   819   (in mm to 999 and
               400 "   "   "   984   thereafter in m)
               500 "   "   "   1.230
               600 "   "   "   1.475

Run of Rafter	0.1	0.2	0.3	0.4	0.5	0.6	0.7	0.8	0.9	1.0
Length of Rafter	0.246	0.492	0.738	0.983	1.229	1.475	1.721	1.967	2.213	2.459
Length of Hip	0.265	0.531	0.796	1.062	1.327	1.593	1.858	2.123	2.389	2.654

RISE OF COMMON RAFTER 2.356 m PER METRE OF RUN   PITCH 67°

BEVELS: COMMON RAFTER   – SEAT   67
          "   "   – RIDGE 23
         HIP OR VALLEY   – SEAT   59
          "   "   "   – RIDGE 31
         JACK RAFTER   – EDGE   21.5
         PURLIN   – EDGE   68.5
          "   – SIDE   47.5

JACK RAFTERS 333 mm **CENTRES DECREASE**   852   (in mm to 999 and
                400 "   "   "   1.024   thereafter in m)
                500 "   "   "   1.280
                600 "   "   "   1.535

Run of Rafter	0.1	0.2	0.3	0.4	0.5	0.6	0.7	0.8	0.9	1.0
Length of Rafter	0.256	0.512	0.768	1.024	1.28	1.536	1.792	2.047	2.303	2.559
Length of Hip	0.275	0.55	0.824	1.099	1.374	1.649	1.922	2.198	2.473	2.748

RISE OF COMMON RAFTER 2.475 m **PER METRE OF RUN**   **PITCH** 68°

BEVELS:	COMMON RAFTER	– SEAT	68	
	" "	– RIDGE	22	
	HIP OR VALLEY	– SEAT	60.5	
	" " "	– RIDGE	29.5	
	JACK RAFTER	– EDGE	20.5	
	PURLIN	– EDGE	69.5	
	"	– SIDE	47	

JACK RAFTERS 333 mm **CENTRES DECREASE** 889 (in mm to 999 and
400 " " " 1.068 thereafter in m)
500 " " " 1.335
600 " " " 1.601

Run of Rafter	0.1	0.2	0.3	0.4	0.5	0.6	0.7	0.8	0.9	1.0
Length of Rafter	0.267	0.534	0.801	1.068	1.335	1.602	1.869	2.136	2.403	2.669
Length of Hip	0.285	0.57	0.855	1.14	1.425	1.71	1.995	2.28	2.566	2.851

RISE OF COMMON RAFTER 2.605 m PER METRE OF RUN   PITCH 69°

BEVELS:

COMMON RAFTER		–	SEAT	69
"	"	–	RIDGE	21
HIP OR VALLEY		–	SEAT	61.5
"	" "	–	RIDGE	28.5
JACK RAFTER		–	EDGE	19.5
PURLIN		–	EDGE	70.5
"		–	SIDE	47

JACK RAFTERS 333 mm CENTRES DECREASE 930 (in mm to 999 and thereafter in m)

400 "	"	"	1.116
500 "	"	"	1.395
600 "	"	"	1.674

	0.1	0.2	0.3	0.4	0.5	0.6	0.7	0.8	0.9	1.0
Run of Rafter	0.1	0.2	0.3	0.4	0.5	0.6	0.7	0.8	0.9	1.0
Length of Rafter	0.279	0.558	0.837	1.116	1.395	1.674	1.953	2.232	2.511	2.79
Length of Hip	0.296	0.593	0.889	1.186	1.482	1.779	2.075	2.371	2.688	2.964

RISE OF COMMON RAFTER 2.747 m **PER METRE OF RUN**   **PITCH** 70°

BEVELS: COMMON RAFTER   – SEAT   70
                    ''   ''   – RIDGE 20
              HIP OR VALLEY   – SEAT   63
              ''   ''   ''   – RIDGE 27
              JACK RAFTER   – EDGE   19
              PURLIN   – EDGE   71
                    ''   – SIDE   47

JACK RAFTERS 333 mm **CENTRES DECREASE**   975   (in mm to 999 and
                        400 ''   ''   ''   1.170   thereafter in m)
                        500 ''   ''   ''   1.462
                        600 ''   ''   ''   1.754

Run of Rafter	0.1	0.2	0.3	0.4	0.5	0.6	0.7	0.8	0.9	1.0
Length of Rafter	0.292	0.585	0.877	1.17	1.462	1.754	2.047	2.339	2.631	2.924
Length of Hip	0.309	0.618	0.927	1.236	1.545	1.854	2.163	2.472	2.781	3.09

RISE OF COMMON RAFTER 2.904 m **PER METRE OF RUN**   **PITCH** 71°

BEVELS: COMMON RAFTER   – SEAT   71
                      ″   ″   – RIDGE   19
          HIP OR VALLEY   – SEAT   64
                  ″   ″   ″   – RIDGE   26
          JACK RAFTER   – EDGE   18
          PURLIN   – EDGE   72
                  ″   – SIDE   46.5

JACK RAFTERS 333 mm **CENTRES DECREASE** 1.024   (in mm to 999 and
                 400 ″   ″   ″   1.229   thereafter in m)
                 500 ″   ″   ″   1.536
                 600 ″   ″   ″   1.843

Run of Rafter	0.1	0.2	0.3	0.4	0.5	0.6	0.7	0.8	0.9	1.0
Length of Rafter	0.307	0.614	0.921	1.229	1.536	1.843	2.15	2.457	2.764	3.072
Length of Hip	0.323	0.646	0.969	1.292	1.615	1.938	2.261	2.584	2.907	3.23

RISE OF COMMON RAFTER 3.078 m **PER METRE OF RUN**   **PITCH** 72°

BEVELS: COMMON RAFTER   – SEAT   72
                "   "   – RIDGE   18
            HIP OR VALLEY   – SEAT   65.5
            "   "   "   – RIDGE   24.5
            JACK RAFTER   – EDGE   17
            PURLIN   – EDGE   73
            "   – SIDE   46.5

JACK RAFTERS   333 mm **CENTRES DECREASE**   1.078   (in mm to 999 and
                   400 "   "   "   1.294   thereafter in m)
                   500 "   "   "   1.618
                   600 "   "   "   1.942

Run of Rafter	0.1	0.2	0.3	0.4	0.5	0.6	0.7	0.8	0.9	1.0
Length of Rafter	0.324	0.647	0.971	1.294	1.618	1.942	2.266	2.589	2.912	3.236
Length of Hip	0.339	0.677	1.016	1.355	1.694	2.032	2.371	2.71	3.048	3.387

RISE OF COMMON RAFTER 3.271 m **PER METRE OF RUN**   **PITCH** 73°

	BEVELS:	COMMON RAFTER	– SEAT	73
		" "	– RIDGE	17
		HIP OR VALLEY	– SEAT	66.5
		" " "	– RIDGE	23.5
		JACK RAFTER	– EDGE	16.5
		PURLIN	– EDGE	73.5
		"	– SIDE	46.5

JACK RAFTERS 333 mm **CENTRES DECREASE** 1.140 (in mm to 999 and thereafter in m)

400 "	"	"	1.368
500 "	"	"	1.710
600 "	"	"	2.052

Run of Rafter	0.1	0.2	0.3	0.4	0.5	0.6	0.7	0.8	0.9	1.0
Length of Rafter	0.342	0.684	1.026	1.368	1.71	2.052	2.394	2.736	3.078	3.42
Length of Hip	0.356	0.713	1.069	1.425	1.782	2.138	2.494	2.851	3.207	3.563

RISE OF COMMON RAFTER 3.487 m **PER METRE OF RUN** **PITCH** 74°

BEVELS:	COMMON RAFTER	– SEAT	74
	" "	– RIDGE	16
	HIP OR VALLEY	– SEAT	68
	" " "	– RIDGE	22
	JACK RAFTER	– EDGE	15.5
	PURLIN	– EDGE	74.5
	"	– SIDE	46

JACK RAFTERS 333 mm **CENTRES DECREASE** 1.209 (in mm to 999 and thereafter in m)

	400 "	"	"	1.451
	500 "	"	"	1.814
	600 "	"	"	2.177

Run of Rafter	0.1	0.2	0.3	0.4	0.5	0.6	0.7	0.8	0.9	1.0
Length of Rafter	0.363	0.726	1.088	1.451	1.814	2.177	2.54	2.902	3.265	3.628
Length of Hip	0.376	0.753	1.129	1.505	1.882	2.258	2.634	3.011	3.387	3.763

RISE OF COMMON RAFTER 3.732 m **PER METRE OF RUN**   **PITCH** 75°

BEVELS:	COMMON RAFTER	– SEAT	75
	" "	– RIDGE	15
	HIP OR VALLEY	– SEAT	69
	" " "	– RIDGE	21
	JACK RAFTER	– EDGE	14.5
	PURLIN	– EDGE	75.5
	"	– SIDE	46

JACK RAFTERS	333 mm	**CENTRES DECREASE**		1.289	(in mm to 999 and
	400 "	"	"	1.546	thereafter in m)
	500 "	"	"	1.932	
	600 "	"	"	2.318	

Run of Rafter	0.1	0.2	0.3	0.4	0.5	0.6	0.7	0.8	0.9	1.0
Length of Rafter	0.386	0.773	1.159	1.545	1.932	2.318	2.705	3.091	3.477	3.864
Length of Hip	0.399	0.789	1.197	1.596	1.996	2.395	2.794	3.193	3.592	3.991

5   IMPERIAL CALCULATION TABLES

5° PITCH

RISE OF COMMON RAFTER $1\frac{1}{16}''$ PER FOOT OF RUN

BEVELS :
1. COMMON RAFTER   – SEAT   5
2.   "   "   – RIDGE   85
3. HIP OR VALLEY   – SEAT
4.   "   "   "   – RIDGE
5. JACK RAFTER   – EDGE
6. PURLIN   – EDGE
7.   "   – SIDE

JACK RAFTERS 16 in. CENTRES DECREASE

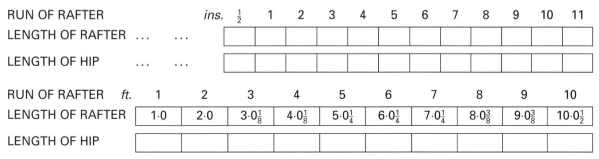

RUN OF RAFTER	*ins.*	$\frac{1}{2}$	1	2	3	4	5	6	7	8	9	10	11
LENGTH OF RAFTER … …													
LENGTH OF HIP … …													

RUN OF RAFTER	*ft.*	1	2	3	4	5	6	7	8	9	10
LENGTH OF RAFTER		1·0	2·0	$3·0\frac{1}{8}$	$4·0\frac{1}{8}$	$5·0\frac{1}{4}$	$6·0\frac{1}{4}$	$7·0\frac{1}{4}$	$8·0\frac{3}{8}$	$9·0\frac{3}{8}$	$10·0\frac{1}{2}$
LENGTH OF HIP											

6° PITCH

RISE OF COMMON RAFTER 1$\frac{1}{4}''$ PER FOOT OF RUN

BEVELS:
1. COMMON RAFTER – SEAT 6
2.   "   "   – RIDGE 84
3. HIP OR VALLEY – SEAT
4. "   "   " – RIDGE
5. JACK RAFTER – EDGE
6. PURLIN – EDGE
7.   "   – SIDE

JACK RAFTERS 16 in. CENTRES DECREASE

RUN OF RAFTER	*ins.*	$\frac{1}{2}$	1	2	3	4	5	6	7	8	9	10	11
LENGTH OF RAFTER													
LENGTH OF HIP													

RUN OF RAFTER *ft.*	1	2	3	4	5	6	7	8	9	10
LENGTH OF RAFTER	1·0	2·0$\frac{1}{8}$	3·0$\frac{1}{8}$	4·0$\frac{1}{4}$	5·0$\frac{1}{4}$	6·0$\frac{3}{8}$	7·0$\frac{1}{2}$	8·0$\frac{1}{2}$	9·0$\frac{5}{8}$	10·0$\frac{5}{8}$
LENGTH OF HIP										

7° PITCH

RISE OF COMMON RAFTER $1\frac{1}{2}''$ PER FOOT OF RUN

BEVELS:
1. COMMON RAFTER   – SEAT   7
2.   "   "   – RIDGE   83
3. HIP OR VALLEY   – SEAT
4.   "   "   "   – RIDGE
5. JACK RAFTER   – EDGE
6. PURLIN   – EDGE
7.   "   – SIDE

JACK RAFTERS 16 in. CENTRES DECREASE

RUN OF RAFTER	ins.	$\frac{1}{2}$	1	2	3	4	5	6	7	8	9	10	11
LENGTH OF RAFTER … …													
LENGTH OF HIP … …													

RUN OF RAFTER	ft.	1	2	3	4	5	6	7	8	9	10
LENGTH OF RAFTER		1·0	$2·0\frac{1}{8}$	$3·0\frac{1}{4}$	$4·0\frac{1}{4}$	$5·0\frac{3}{8}$	$6·0\frac{1}{2}$	$7·0\frac{5}{8}$	$8·0\frac{5}{8}$	$9·0\frac{3}{4}$	$10·0\frac{7}{8}$
LENGTH OF HIP											

8° PITCH

RISE OF COMMON RAFTER $1\frac{11}{16}''$ PER FOOT OF RUN

BEVELS :
1. COMMON RAFTER – SEAT 8
2. " " – RIDGE 82
3. HIP OR VALLEY – SEAT
4. " " " – RIDGE
5. JACK RAFTER – EDGE
6. PURLIN – EDGE
7. " – SIDE

JACK RAFTERS 16 in. CENTRES DECREASE

RUN OF RAFTER *ins.*	$\frac{1}{2}$	1	2	3	4	5	6	7	8	9	10	11
LENGTH OF RAFTER												
LENGTH OF HIP												

RUN OF RAFTER *ft.*	1	2	3	4	5	6	7	8	9	10
LENGTH OF RAFTER	$1 \cdot 0\frac{1}{8}$	$2 \cdot 0\frac{1}{4}$	$3 \cdot 0\frac{3}{8}$	$4 \cdot 0\frac{1}{2}$	$5 \cdot 0\frac{5}{8}$	$6 \cdot 0\frac{3}{4}$	$7 \cdot 0\frac{7}{8}$	$8 \cdot 1$	$9 \cdot 1$	$10 \cdot 1\frac{1}{8}$
LENGTH OF HIP										

9° PITCH

RISE OF COMMON RAFTER $1\frac{7}{8}''$ PER FOOT OF RUN

BEVELS:
1. COMMON RAFTER   – SEAT   9
2.   ''   ''   – RIDGE 81
3. HIP OR VALLEY   – SEAT
4.   ''   ''   ''   – RIDGE
5. JACK RAFTER   – EDGE
6. PURLIN   – EDGE
7.   ''   – SIDE

JACK RAFTERS 16 in. CENTRES DECREASE

RUN OF RAFTER *ins.*	$\frac{1}{2}$	1	2	3	4	5	6	7	8	9	10	11
LENGTH OF RAFTER … …												
LENGTH OF HIP … …												

RUN OF RAFTER *ft.*	1	2	3	4	5	6	7	8	9	10
LENGTH OF RAFTER	$1 \cdot 0\frac{1}{8}$	$2 \cdot 0\frac{1}{4}$	$3 \cdot 0\frac{3}{8}$	$4 \cdot 0\frac{5}{8}$	$5 \cdot 0\frac{3}{4}$	$6 \cdot 0\frac{7}{8}$	$7 \cdot 1$	$8 \cdot 1\frac{1}{4}$	$9 \cdot 1\frac{3}{8}$	$10 \cdot 1\frac{1}{2}$
LENGTH OF HIP										

10° PITCH

RISE OF COMMON RAFTER $2\frac{1}{8}''$ PER FOOT OF RUN

BEVELS:
1. COMMON RAFTER   – SEAT   10
2.   "   "   – RIDGE   80
3. HIP OR VALLEY   – SEAT
4.   "   "   "   – RIDGE
5. JACK RAFTER   – EDGE
6. PURLIN   – EDGE
7.   "   – SIDE

JACK RAFTERS 16 in. CENTRES DECREASE

RUN OF RAFTER		ins.	$\frac{1}{2}$	1	2	3	4	5	6	7	8	9	10	11
LENGTH OF RAFTER														
LENGTH OF HIP														

RUN OF RAFTER ft.	1	2	3	4	5	6	7	8	9	10
LENGTH OF RAFTER	$1{\cdot}0\frac{1}{8}$	$2{\cdot}0\frac{3}{8}$	$3{\cdot}0\frac{1}{2}$	$4{\cdot}0\frac{3}{4}$	$5{\cdot}1$	$6{\cdot}1\frac{1}{8}$	$7{\cdot}1\frac{1}{4}$	$8{\cdot}1\frac{1}{2}$	$9{\cdot}1\frac{5}{8}$	$10{\cdot}1\frac{7}{8}$
LENGTH OF HIP										

11° PITCH

RISE OF COMMON RAFTER $2\frac{5}{16}''$ PER FOOT OF RUN

BEVELS:
1. COMMON RAFTER – SEAT 11
2.   "   "   – RIDGE 79
3. HIP OR VALLEY – SEAT
4. "   "   "   – RIDGE
5. JACK RAFTER – EDGE
6. PURLIN – EDGE
7.   "   – SIDE

JACK RAFTERS 16 in. CENTRES DECREASE

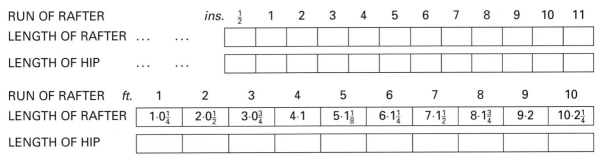

RUN OF RAFTER	ins.	$\frac{1}{2}$	1	2	3	4	5	6	7	8	9	10	11
LENGTH OF RAFTER													
LENGTH OF HIP													

RUN OF RAFTER	ft.	1	2	3	4	5	6	7	8	9	10
LENGTH OF RAFTER		$1 \cdot 0\frac{1}{4}$	$2 \cdot 0\frac{1}{2}$	$3 \cdot 0\frac{3}{4}$	$4 \cdot 1$	$5 \cdot 1\frac{1}{8}$	$6 \cdot 1\frac{1}{4}$	$7 \cdot 1\frac{1}{2}$	$8 \cdot 1\frac{3}{4}$	$9 \cdot 2$	$10 \cdot 2\frac{1}{4}$
LENGTH OF HIP											

12° PITCH

RISE OF COMMON RAFTER 2$\frac{9}{16}$" PER FOOT OF RUN

BEVELS :   1. COMMON RAFTER   – SEAT   12
                2.   "   "   – RIDGE 78
                3. HIP OR VALLEY   – SEAT
                4.   "   "   "   – RIDGE
                5. JACK RAFTER   – EDGE
                6. PURLIN   – EDGE
                7.   "   – SIDE

JACK RAFTERS 16 in. CENTRES DECREASE

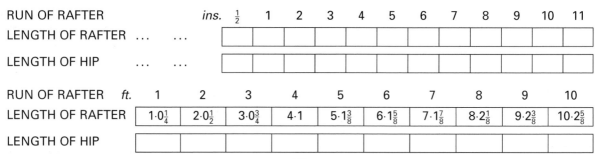

RUN OF RAFTER	ins.	$\frac{1}{2}$	1	2	3	4	5	6	7	8	9	10	11
LENGTH OF RAFTER													
LENGTH OF HIP													

RUN OF RAFTER	ft.	1	2	3	4	5	6	7	8	9	10
LENGTH OF RAFTER		1·0$\frac{1}{4}$	2·0$\frac{1}{2}$	3·0$\frac{3}{4}$	4·1	5·1$\frac{3}{8}$	6·1$\frac{5}{8}$	7·1$\frac{7}{8}$	8·2$\frac{1}{8}$	9·2$\frac{3}{8}$	10·2$\frac{5}{8}$
LENGTH OF HIP											

13° PITCH

RISE OF COMMON RAFTER $2\frac{3}{4}''$ PER FOOT OF RUN

BEVELS :
1. COMMON RAFTER   – SEAT 13
2.   "   "   – RIDGE 77
3. HIP OR VALLEY   – SEAT
4.   "   "   "   – RIDGE
5. JACK RAFTER   – EDGE
6. PURLIN   – EDGE
7.   "   – SIDE

JACK RAFTERS 16 in. CENTRES DECREASE

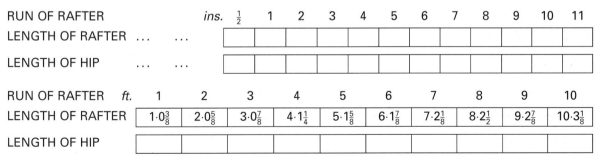

RUN OF RAFTER	*ins.*	$\frac{1}{2}$	1	2	3	4	5	6	7	8	9	10	11
LENGTH OF RAFTER													
LENGTH OF HIP													

RUN OF RAFTER	*ft.*	1	2	3	4	5	6	7	8	9	10
LENGTH OF RAFTER		$1 \cdot 0\frac{3}{8}$	$2 \cdot 0\frac{5}{8}$	$3 \cdot 0\frac{7}{8}$	$4 \cdot 1\frac{1}{4}$	$5 \cdot 1\frac{5}{8}$	$6 \cdot 1\frac{7}{8}$	$7 \cdot 2\frac{1}{8}$	$8 \cdot 2\frac{1}{2}$	$9 \cdot 2\frac{7}{8}$	$10 \cdot 3\frac{1}{8}$
LENGTH OF HIP											

14° PITCH

RISE OF COMMON RAFTER 3″ PER FOOT OF RUN

BEVELS :
1. COMMON RAFTER – SEAT 14
2. ″ ″ – RIDGE 76
3. HIP OR VALLEY – SEAT
4. ″ ″ ″ – RIDGE
5. JACK RAFTER – EDGE
6. PURLIN – EDGE
7. ″ – SIDE

JACK RAFTERS 16 in. CENTRES DECREASE

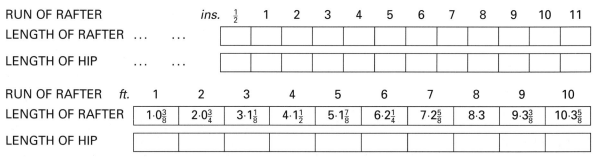

RUN OF RAFTER	ins.	$\frac{1}{2}$	1	2	3	4	5	6	7	8	9	10	11
LENGTH OF RAFTER													
LENGTH OF HIP													

RUN OF RAFTER	ft.	1	2	3	4	5	6	7	8	9	10
LENGTH OF RAFTER		$1\cdot0\frac{3}{8}$	$2\cdot0\frac{3}{4}$	$3\cdot1\frac{1}{8}$	$4\cdot1\frac{1}{2}$	$5\cdot1\frac{7}{8}$	$6\cdot2\frac{1}{4}$	$7\cdot2\frac{5}{8}$	$8\cdot3$	$9\cdot3\frac{3}{8}$	$10\cdot3\frac{5}{8}$
LENGTH OF HIP											

15° PITCH

RISE OF COMMON RAFTER $3\frac{3}{16}''$ PER FOOT OF RUN

BEVELS:
1. COMMON RAFTER   – SEAT   15
2.   "   "   – RIDGE 75
3. HIP OR VALLEY   – SEAT
4.   "   "   "   – RIDGE
5. JACK RAFTER   – EDGE
6. PURLIN   – EDGE
7.   "   – SIDE

JACK RAFTERS 16 in. CENTRES DECREASE

RUN OF RAFTER	ins.	$\frac{1}{2}$	1	2	3	4	5	6	7	8	9	10	11
LENGTH OF RAFTER													
LENGTH OF HIP													

RUN OF RAFTER	ft.	1	2	3	4	5	6	7	8	9	10
LENGTH OF RAFTER		$1 \cdot 0\frac{3}{8}$	$2 \cdot 0\frac{7}{8}$	$3 \cdot 1\frac{1}{4}$	$4 \cdot 1\frac{5}{8}$	$5 \cdot 2\frac{1}{8}$	$6 \cdot 2\frac{1}{2}$	$7 \cdot 3$	$8 \cdot 3\frac{3}{8}$	$9 \cdot 3\frac{3}{4}$	$10 \cdot 4\frac{1}{4}$
LENGTH OF HIP											

16° or GRECIAN PITCH

RISE OF COMMON RAFTER $3\frac{7}{16}''$ PER FOOT OF RUN

BEVELS:
1. COMMON RAFTER – SEAT 16
2.   "   " – RIDGE 74
3. HIP OR VALLEY – SEAT $11\frac{1}{2}$
4.   "   "   " – RIDGE $78\frac{1}{2}$
5. JACK RAFTER – EDGE 44
6. PURLIN – EDGE 46
7.   " – SIDE $74\frac{1}{2}$

JACK RAFTERS 16 in. CENTRES DECREASE $16\frac{5}{8}''$, 18 in.—$18\frac{3}{4}''$, 24 in.—$2'1''$

RUN OF RAFTER	ins.	$\frac{1}{2}$	1	2	3	4	5	6	7	8	9	10	11
LENGTH OF RAFTER		$\frac{1}{2}$	1	$2\frac{1}{8}$	$3\frac{1}{8}$	$4\frac{1}{8}$	$5\frac{1}{4}$	$6\frac{1}{4}$	$7\frac{1}{4}$	$8\frac{3}{8}$	$9\frac{3}{8}$	$10\frac{3}{8}$	$11\frac{1}{2}$
LENGTH OF HIP		$\frac{3}{4}$	$1\frac{1}{2}$	$2\frac{7}{8}$	$4\frac{3}{8}$	$5\frac{3}{4}$	$7\frac{1}{4}$	$8\frac{5}{8}$	$10\frac{1}{8}$	$11\frac{1}{2}$	13	$14\frac{3}{8}$	$15\frac{7}{8}$

RUN OF RAFTER	ft.	1	2	3	4	5	6	7	8	9	10
LENGTH OF RAFTER		$1 \cdot 0\frac{1}{2}$	$2 \cdot 1$	$3 \cdot 1\frac{1}{2}$	$4 \cdot 1\frac{7}{8}$	$5 \cdot 2\frac{3}{8}$	$6 \cdot 2\frac{7}{8}$	$7 \cdot 3\frac{3}{8}$	$8 \cdot 3\frac{7}{8}$	$9 \cdot 4\frac{7}{8}$	$10 \cdot 4\frac{7}{8}$
LENGTH OF HIP		$1 \cdot 5\frac{1}{4}$	$2 \cdot 10\frac{5}{8}$	$4 \cdot 4$	$5 \cdot 9\frac{1}{4}$	$7 \cdot 2\frac{1}{2}$	$8 \cdot 7\frac{7}{8}$	$10 \cdot 1\frac{1}{8}$	$11 \cdot 6\frac{1}{2}$	$12 \cdot 11\frac{3}{4}$	$14 \cdot 5\frac{1}{8}$

17° PITCH

RISE OF COMMON RAFTER $3\frac{11}{16}''$ PER FOOT OF RUN

BEVELS:
1. COMMON RAFTER   – SEAT   17
2.   ''   ''   – RIDGE 73
3. HIP OR VALLEY   – SEAT   12
4.   ''   ''   ''   – RIDGE 78
5. JACK RAFTER   – EDGE $43\frac{1}{2}$
6. PURLIN   – EDGE $46\frac{1}{2}$
7.   ''   – SIDE $73\frac{1}{2}$

JACK RAFTERS 16 in. CENTRES DECREASE $16\frac{3}{4}''$, 18 in.—$18\frac{7}{8}''$, 24 in.—2' $1\frac{1}{8}''$

RUN OF RAFTER	*ins.*	$\frac{1}{2}$	1	2	3	4	5	6	7	8	9	10	11
LENGTH OF RAFTER		$\frac{1}{2}$	1	$2\frac{1}{8}$	$3\frac{1}{8}$	$4\frac{1}{4}$	$5\frac{1}{4}$	$6\frac{1}{4}$	$7\frac{3}{8}$	$8\frac{3}{8}$	$9\frac{3}{8}$	$10\frac{1}{2}$	$11\frac{1}{2}$
LENGTH OF HIP		$\frac{3}{4}$	$1\frac{1}{2}$	$2\frac{7}{8}$	$4\frac{3}{8}$	$5\frac{3}{4}$	$7\frac{1}{4}$	$8\frac{5}{8}$	$10\frac{1}{8}$	$11\frac{1}{2}$	13	$14\frac{1}{2}$	$15\frac{7}{8}$

RUN OF RAFTER	*ft.*	1	2	3	4	5	6	7	8	9	10
LENGTH OF RAFTER		$1\cdot0\frac{1}{2}$	$2\cdot1\frac{1}{8}$	$3\cdot1\frac{5}{8}$	$4\cdot2\frac{1}{4}$	$5\cdot2\frac{3}{4}$	$6\cdot3\frac{1}{4}$	$7\cdot3\frac{7}{8}$	$8\cdot4\frac{3}{8}$	$9\cdot4\frac{7}{8}$	$10\cdot5\frac{1}{2}$
LENGTH OF HIP		$1\cdot5\frac{3}{8}$	$2\cdot10\frac{3}{4}$	$4\cdot4\frac{1}{8}$	$5\cdot9\frac{3}{8}$	$7\cdot2\frac{3}{4}$	$8\cdot8\frac{1}{8}$	$10\cdot1\frac{1}{2}$	$11\cdot6\frac{7}{8}$	$13\cdot0\frac{1}{4}$	$14\cdot5\frac{1}{2}$

$17\frac{1}{2}°$ PITCH

RISE OF COMMON RAFTER $3\frac{3}{4}''$ PER FOOT OF RUN

BEVELS :
1. COMMON RAFTER – SEAT $17\frac{1}{2}$
2.   "   "   – RIDGE $72\frac{1}{2}$
3. HIP OR VALLEY – SEAT $12\frac{1}{2}$
4.   "   "   "   – RIDGE $77\frac{1}{2}$
5. JACK RAFTER – EDGE $43\frac{1}{2}$
6. PURLIN – EDGE $46\frac{1}{2}$
7.   "   – SIDE $73\frac{1}{2}$

JACK RAFTERS 16 in. CENTRES DECREASE $16\frac{3}{4}''$, 18 in.—$18\frac{7}{8}''$, 24 in.—$2'\ 1\frac{1}{8}''$

RUN OF RAFTER	ins.	$\frac{1}{2}$	1	2	3	4	5	6	7	8	9	10	11
LENGTH OF RAFTER		$\frac{1}{2}$	1	$2\frac{1}{8}$	$3\frac{1}{8}$	$4\frac{1}{4}$	$5\frac{1}{4}$	$6\frac{1}{4}$	$7\frac{3}{8}$	$8\frac{3}{8}$	$9\frac{3}{8}$	$10\frac{1}{2}$	$11\frac{1}{2}$
LENGTH OF HIP		$\frac{3}{4}$	$1\frac{1}{2}$	$2\frac{7}{8}$	$4\frac{3}{8}$	$5\frac{3}{4}$	$7\frac{1}{4}$	$8\frac{5}{8}$	$10\frac{1}{8}$	$11\frac{1}{2}$	13	$14\frac{1}{2}$	$15\frac{7}{8}$

RUN OF RAFTER	ft.	1	2	3	4	5	6	7	8	9	10
LENGTH OF RAFTER		$1·0\frac{5}{8}$	$2·1\frac{1}{8}$	$3·1\frac{3}{4}$	$4·2\frac{3}{8}$	5·3	$6·3\frac{1}{2}$	7·4	$8·4\frac{5}{8}$	$9·5\frac{1}{4}$	$10·5\frac{7}{8}$
LENGTH OF HIP		$1·5\frac{3}{8}$	$2·10\frac{7}{8}$	$4·4\frac{1}{4}$	$5·9\frac{1}{2}$	7·3	$8·8\frac{1}{4}$	$10·1\frac{5}{8}$	11·7	$13·0\frac{1}{2}$	14·6

18° PITCH

RISE OF COMMON RAFTER $3\frac{7}{8}''$ PER FOOT OF RUN

BEVELS:
1. COMMON RAFTER   – SEAT   18
2.   ''   ''   – RIDGE   72
3. HIP OR VALLEY   – SEAT   13
4.   ''   ''   ''   – RIDGE   77
5. JACK RAFTER   – EDGE $43\frac{1}{2}$
6. PURLIN   – EDGE $46\frac{1}{2}$
7.   ''   – SIDE   73

JACK RAFTERS 16 in. CENTRES DECREASE $16\frac{7}{8}''$, 18 in.—$18\frac{7}{8}''$, 24 in.—2' $1\frac{1}{4}''$

RUN OF RAFTER *ins.*	$\frac{1}{2}$	1	2	3	4	5	6	7	8	9	10	11
LENGTH OF RAFTER	$\frac{1}{2}$	1	$2\frac{1}{8}$	$3\frac{1}{8}$	$4\frac{1}{4}$	$5\frac{1}{4}$	$6\frac{3}{8}$	$7\frac{3}{8}$	$8\frac{3}{8}$	$9\frac{1}{2}$	$10\frac{1}{2}$	$11\frac{5}{8}$
LENGTH OF HIP	$\frac{3}{4}$	$1\frac{1}{2}$	$2\frac{7}{8}$	$4\frac{3}{8}$	$5\frac{3}{4}$	$7\frac{1}{4}$	$8\frac{3}{4}$	$10\frac{1}{8}$	$11\frac{5}{8}$	$13\frac{1}{8}$	$14\frac{1}{2}$	16

RUN OF RAFTER *ft.*	1	2	3	4	5	6	7	8	9	10
LENGTH OF RAFTER	$1\cdot0\frac{5}{8}$	$2\cdot1\frac{1}{4}$	$3\cdot1\frac{7}{8}$	$4\cdot2\frac{1}{2}$	$5\cdot3\frac{1}{8}$	$6\cdot3\frac{3}{4}$	$7\cdot4\frac{3}{8}$	$8\cdot5$	$9\cdot5\frac{1}{2}$	$10\cdot6\frac{1}{8}$
LENGTH OF HIP	$1\cdot5\frac{3}{8}$	$2\cdot10\frac{7}{8}$	$4\cdot4\frac{1}{4}$	$5\cdot9\frac{5}{8}$	$7\cdot3\frac{1}{8}$	$8\cdot8\frac{1}{2}$	$10\cdot2$	$11\cdot7\frac{3}{8}$	$13\cdot0\frac{3}{4}$	$14\cdot6\frac{1}{4}$

19° PITCH

RISE OF COMMON RAFTER 4$\frac{1}{8}$" PER FOOT OF RUN

BEVELS :
1. COMMON RAFTER   – SEAT   19
2.   "   "   – RIDGE   71
3. HIP OR VALLEY   – SEAT   13$\frac{1}{2}$
4.   "   "   "   – RIDGE   76$\frac{1}{2}$
5. JACK RAFTER   – EDGE   43$\frac{1}{2}$
6. PURLIN   – EDGE   46$\frac{1}{2}$
7.   "   – SIDE   72

JACK RAFTERS 16 in. CENTRES DECREASE 16$\frac{7}{8}$", 18 in.—19", 24 in.—2' 1$\frac{3}{8}$"

RUN OF RAFTER	ins.	$\frac{1}{2}$	1	2	3	4	5	6	7	8	9	10	11
LENGTH OF RAFTER		$\frac{1}{2}$	1	2$\frac{1}{8}$	3$\frac{1}{8}$	4$\frac{1}{4}$	5$\frac{1}{4}$	6$\frac{3}{8}$	7$\frac{3}{8}$	8$\frac{1}{2}$	9$\frac{1}{2}$	10$\frac{5}{8}$	11$\frac{5}{8}$
LENGTH OF HIP		$\frac{3}{4}$	1$\frac{1}{2}$	2$\frac{7}{8}$	4$\frac{3}{8}$	5$\frac{7}{8}$	7$\frac{1}{4}$	8$\frac{3}{4}$	10$\frac{1}{8}$	11$\frac{5}{8}$	13$\frac{1}{8}$	14$\frac{1}{2}$	16

RUN OF RAFTER	ft.	1	2	3	4	5	6	7	8	9	10
LENGTH OF RAFTER		1·0$\frac{3}{4}$	2·1$\frac{3}{8}$	3·2$\frac{1}{8}$	4·2$\frac{3}{4}$	5·3$\frac{1}{2}$	6·4$\frac{1}{8}$	7·4$\frac{7}{8}$	8·5$\frac{1}{2}$	9·6$\frac{1}{4}$	10·6$\frac{7}{8}$
LENGTH OF HIP		1·5$\frac{1}{2}$	2·10$\frac{7}{8}$	4·4$\frac{3}{8}$	5·9$\frac{7}{8}$	7·3$\frac{1}{4}$	8·8$\frac{3}{4}$	10·2$\frac{1}{4}$	11·7$\frac{5}{8}$	13·1$\frac{1}{8}$	14·6$\frac{1}{2}$

20° PITCH

RISE OF COMMON RAFTER $4\frac{3}{8}''$ PER FOOT OF RUN

BEVELS:
1. COMMON RAFTER   – SEAT   20
2.   ″   ″   – RIDGE   70
3. HIP OR VALLEY   – SEAT $14\frac{1}{2}$
4.   ″   ″   ″   – RIDGE $75\frac{1}{2}$
5. JACK RAFTER   – EDGE   43
6. PURLIN   – EDGE   47
7.   ″   – SIDE   71

JACK RAFTERS 16 in. CENTRES DECREASE 17″, 18 in.—$19\frac{1}{8}''$, 24 in.—2′ $1\frac{1}{2}''$

RUN OF RAFTER	*ins.*	$\frac{1}{2}$	1	2	3	4	5	6	7	8	9	10	11
LENGTH OF RAFTER		$\frac{1}{2}$	$1\frac{1}{8}$	$2\frac{1}{8}$	$3\frac{1}{4}$	$4\frac{1}{4}$	$5\frac{3}{8}$	$6\frac{3}{8}$	$7\frac{1}{2}$	$8\frac{1}{2}$	$9\frac{5}{8}$	$10\frac{5}{8}$	$11\frac{3}{4}$
LENGTH OF HIP		$\frac{3}{4}$	$1\frac{1}{2}$	$2\frac{7}{8}$	$4\frac{3}{8}$	$5\frac{7}{8}$	$7\frac{1}{4}$	$8\frac{3}{4}$	$10\frac{1}{4}$	$11\frac{5}{8}$	$13\frac{1}{8}$	$14\frac{5}{8}$	16

RUN OF RAFTER	*ft.*	1	2	3	4	5	6	7	8	9	10
LENGTH OF RAFTER		$1 \cdot 0\frac{3}{4}$	$2 \cdot 1\frac{1}{2}$	$3 \cdot 2\frac{1}{4}$	$4 \cdot 3\frac{1}{8}$	$5 \cdot 3\frac{7}{8}$	$6 \cdot 4\frac{5}{8}$	$7 \cdot 5\frac{3}{8}$	$8 \cdot 6\frac{1}{8}$	$9 \cdot 6\frac{7}{8}$	$10 \cdot 7\frac{3}{4}$
LENGTH OF HIP		$1 \cdot 5\frac{1}{2}$	$2 \cdot 11$	$4 \cdot 4\frac{1}{2}$	$5 \cdot 10\frac{1}{8}$	$7 \cdot 3\frac{5}{8}$	$8 \cdot 9\frac{1}{8}$	$10 \cdot 2\frac{5}{8}$	$11 \cdot 8\frac{1}{8}$	$13 \cdot 1\frac{5}{8}$	$14 \cdot 7\frac{1}{8}$

21° PITCH

RISE OF COMMON RAFTER $4\frac{5}{8}''$ PER FOOT OF RUN

BEVELS:	1. COMMON RAFTER	–	SEAT	21
	2. " "	–	RIDGE	69
	3. HIP OR VALLEY	–	SEAT	15
	4. " " "	–	RIDGE	75
	5. JACK RAFTER	–	EDGE	43
	6. PURLIN	–	EDGE	47
	7. "	–	SIDE	$70\frac{1}{2}$

JACK RAFTERS 16 in. CENTRES DECREASE $17\frac{1}{8}''$, 18 in.—$19\frac{1}{4}''$, 24 in.—2′ $1\frac{3}{4}''$

RUN OF RAFTER *ins.*	$\frac{1}{2}$	1	2	3	4	5	6	7	8	9	10	11
LENGTH OF RAFTER	$\frac{1}{2}$	$1\frac{1}{8}$	$2\frac{1}{8}$	$3\frac{1}{4}$	$4\frac{1}{4}$	$5\frac{3}{8}$	$6\frac{3}{8}$	$7\frac{1}{2}$	$8\frac{5}{8}$	$9\frac{5}{8}$	$10\frac{3}{4}$	$11\frac{3}{4}$
LENGTH OF HIP	$\frac{3}{4}$	$1\frac{1}{2}$	$2\frac{7}{8}$	$4\frac{3}{8}$	$5\frac{7}{8}$	$7\frac{3}{8}$	$8\frac{3}{4}$	$10\frac{1}{4}$	$11\frac{3}{4}$	$13\frac{1}{8}$	$14\frac{5}{8}$	$16\frac{1}{8}$

RUN OF RAFTER *ft.*	1	2	3	4	5	6	7	8	9	10
LENGTH OF RAFTER	$1{\cdot}0\frac{7}{8}$	$2{\cdot}1\frac{3}{4}$	$3{\cdot}2\frac{1}{2}$	$4{\cdot}3\frac{3}{8}$	$5{\cdot}4\frac{1}{4}$	$6{\cdot}5\frac{1}{8}$	7·6	$8{\cdot}6\frac{7}{8}$	$9{\cdot}7\frac{5}{8}$	$10{\cdot}8\frac{1}{2}$
LENGTH OF HIP	$1{\cdot}5\frac{3}{8}$	$2{\cdot}11\frac{1}{8}$	$4{\cdot}4\frac{3}{4}$	$5{\cdot}10\frac{1}{4}$	$7{\cdot}3\frac{7}{8}$	$8{\cdot}9\frac{1}{2}$	10·3	$11{\cdot}8\frac{5}{8}$	$13{\cdot}2\frac{1}{8}$	$14{\cdot}7\frac{3}{4}$

22° PITCH

RISE OF COMMON RAFTER 4⅞″ PER FOOT OF RUN

BEVELS:				
1.	COMMON RAFTER	–	SEAT	22
2.	″ ″	–	RIDGE	68
3.	HIP OR VALLEY	–	SEAT	16
4.	″ ″ ″	–	RIDGE	74
5.	JACK RAFTER	–	EDGE	43
6.	PURLIN	–	EDGE	47
7.	″	–	SIDE	69½

JACK RAFTERS 16 in. CENTRES DECREASE 17¼″, 18 in.—19⅜″, 24 in.—2′ 1⅞″

RUN OF RAFTER *ins.*	½	1	2	3	4	5	6	7	8	9	10	11
LENGTH OF RAFTER	½	1⅛	2⅛	3¼	4⅜	5⅜	6½	7½	8⅝	9¾	10¾	11⅞
LENGTH OF HIP	¾	1½	3	4⅜	5⅞	7⅜	8⅞	10¼	11¾	13¼	14¾	16⅛

RUN OF RAFTER *ft.*	1	2	3	4	5	6	7	8	9	10
LENGTH OF RAFTER	1·1	2·1⅞	3·2⅞	4·3¾	5·4¾	6·5⅝	7·6⅝	8·7½	9·8½	10·9⅜
LENGTH OF HIP	1·5⅝	2·11¼	4·4⅞	5·10⅝	7·4¼	8·9⅞	10·3½	11·9⅛	13·2⅞	14·8½

$22\frac{1}{2}°$ PITCH

RISE OF COMMON RAFTER 5″ PER FOOT OF RUN

BEVELS :
1. COMMON RAFTER – SEAT $22\frac{1}{2}$
2.   ″   ″   – RIDGE $67\frac{1}{2}$
3. HIP OR VALLEY – SEAT $16\frac{1}{4}$
4.   ″   ″   ″   – RIDGE $73\frac{3}{4}$
5. JACK RAFTER – EDGE $42\frac{3}{4}$
6. PURLIN – EDGE $47\frac{1}{2}$
7.   ″   – SIDE 69

JACK RAFTERS 16 in. CENTRES DECREASE $17\frac{1}{4}″$, 18 in.—$19\frac{3}{8}″$, 24 in.—2′ 2″

RUN OF RAFTER	ins.	$\frac{1}{2}$	1	2	3	4	5	6	7	8	9	10	11
LENGTH OF RAFTER		$\frac{1}{2}$	$1\frac{1}{8}$	$2\frac{1}{8}$	$3\frac{1}{4}$	$4\frac{3}{8}$	$5\frac{3}{8}$	$6\frac{1}{2}$	$7\frac{1}{2}$	$8\frac{5}{8}$	$9\frac{3}{4}$	$10\frac{3}{4}$	$11\frac{7}{8}$
LENGTH OF HIP		$\frac{3}{4}$	$1\frac{1}{2}$	3	$4\frac{3}{8}$	$5\frac{7}{8}$	$7\frac{3}{8}$	$8\frac{7}{8}$	$10\frac{1}{4}$	$11\frac{3}{4}$	$13\frac{1}{4}$	$14\frac{3}{4}$	$16\frac{1}{8}$

RUN OF RAFTER	ft.	1	2	3	4	5	6	7	8	9	10
LENGTH OF RAFTER		1·1	2·2	3·3	4·4	5·5	6·6	7·7	8·8	$9·8\frac{7}{8}$	$10·9\frac{7}{8}$
LENGTH OF HIP		$1·5\frac{5}{8}$	$2·11\frac{1}{4}$	4·5	$5·10\frac{3}{4}$	$7·5\frac{7}{8}$	$8·9\frac{1}{2}$	$10·3\frac{3}{4}$	$11·9\frac{1}{2}$	13·3	$14·8\frac{3}{4}$

23° PITCH

RISE OF COMMON RAFTER 5$\frac{1}{8}$″ PER FOOT OF RUN

BEVELS :
1. COMMON RAFTER   – SEAT   23
2.   ″   ″   – RIDGE 67
3. HIP OR VALLEY   – SEAT   16$\frac{1}{2}$
4.   ″   ″   ″   – RIDGE 73$\frac{1}{2}$
5. JACK RAFTER   – EDGE   42$\frac{1}{2}$
6. PURLIN   – EDGE   47$\frac{1}{2}$
7.   ″   – SIDE   68$\frac{1}{2}$

JACK RAFTERS 16 in. CENTRES DECREASE 17$\frac{3}{8}$″, 18 in.—19$\frac{1}{2}$″, 24 in.—2′ 2$\frac{1}{8}$″

RUN OF RAFTER	ins.	$\frac{1}{2}$	1	2	3	4	5	6	7	8	9	10	11
LENGTH OF RAFTER		$\frac{1}{2}$	$1\frac{1}{8}$	$2\frac{1}{8}$	$3\frac{1}{4}$	$4\frac{3}{8}$	$5\frac{5}{8}$	$6\frac{1}{2}$	$7\frac{5}{8}$	$8\frac{3}{4}$	$9\frac{3}{4}$	$10\frac{7}{8}$	12
LENGTH OF HIP		$\frac{3}{4}$	$1\frac{1}{2}$	3	$4\frac{3}{8}$	$5\frac{7}{8}$	$7\frac{3}{8}$	$8\frac{7}{8}$	$10\frac{3}{8}$	$11\frac{3}{4}$	$13\frac{1}{4}$	$14\frac{3}{4}$	$16\frac{1}{4}$

RUN OF RAFTER	ft.	1	2	3	4	5	6	7	8	9	10
LENGTH OF RAFTER		1·1	$2·2\frac{1}{8}$	$3·3\frac{1}{8}$	$4·4\frac{1}{8}$	$5·5\frac{5}{8}$	$6·6\frac{1}{4}$	$7·7\frac{1}{4}$	$8·8\frac{1}{4}$	$9·9\frac{3}{8}$	$10·10\frac{3}{8}$
LENGTH OF HIP		$1·5\frac{3}{4}$	$2·11\frac{3}{8}$	$4·5\frac{1}{8}$	$5·10\frac{7}{8}$	$7·4\frac{1}{2}$	$8·10\frac{1}{4}$	10·4	$11·9\frac{3}{4}$	$13·3\frac{1}{2}$	$14·9\frac{1}{8}$

24° or ROMAN PITCH

RISE OF COMMON RAFTER 5$\frac{5}{16}$″ PER FOOT OF RUN

BEVELS :
1. COMMON RAFTER – SEAT 24
2.   ″   ″ – RIDGE 66
3. HIP OR VALLEY – SEAT 17$\frac{1}{2}$
4.   ″   ″   ″ – RIDGE 72$\frac{1}{2}$
5. JACK RAFTER – EDGE 42$\frac{1}{2}$
6. PURLIN – EDGE 47$\frac{1}{2}$
7.   ″ – SIDE 68

JACK RAFTERS 16 in. CENTRES DECREASE 17$\frac{1}{2}$″, 18 in.—19$\frac{3}{4}$″, 24 in.—2′ 2$\frac{1}{4}$″

RUN OF RAFTER	ins.	$\frac{1}{2}$	1	2	3	4	5	6	7	8	9	10	11
LENGTH OF RAFTER		$\frac{1}{2}$	1$\frac{1}{8}$	2$\frac{1}{8}$	3$\frac{1}{4}$	4$\frac{3}{8}$	5$\frac{1}{2}$	6$\frac{1}{2}$	7$\frac{5}{8}$	8$\frac{3}{4}$	9$\frac{7}{8}$	11	12
LENGTH OF HIP		$\frac{3}{4}$	1$\frac{1}{2}$	3	4$\frac{1}{2}$	5$\frac{7}{8}$	7$\frac{3}{8}$	8$\frac{7}{8}$	10$\frac{3}{8}$	11$\frac{7}{8}$	13$\frac{3}{8}$	14$\frac{7}{8}$	16$\frac{1}{4}$

RUN OF RAFTER	ft.	1	2	3	4	5	6	7	8	9	10
LENGTH OF RAFTER		1·1$\frac{1}{8}$	2·2$\frac{1}{4}$	3·3$\frac{3}{8}$	4·4$\frac{1}{2}$	5·5$\frac{5}{8}$	6·6$\frac{3}{4}$	7·8	8·9$\frac{1}{8}$	9·10$\frac{1}{4}$	10·11$\frac{3}{8}$
LENGTH OF HIP		1·5$\frac{3}{4}$	2·11$\frac{5}{8}$	4·5$\frac{3}{8}$	5·11$\frac{1}{8}$	7·5	8·10$\frac{3}{4}$	10·4$\frac{5}{8}$	11·10$\frac{3}{8}$	13·4$\frac{1}{8}$	14·10

25°

RISE OF COMMON RAFTER 5$\frac{9}{16}$'' PER FOOT OF RUN

BEVELS:
1. COMMON RAFTER   – SEAT   25
2.   ''   ''   – RIDGE 65
3. HIP OR VALLEY   – SEAT   17
4.   ''   ''   ''   – RIDGE 72
5. JACK RAFTER   – EDGE   42
6. PURLIN   – EDGE   48
7.   ''   – SIDE   67

JACK RAFTERS 16 in. CENTRES DECREASE 17$\frac{5}{8}$'', 18 in.—19$\frac{7}{8}$'', 24 in.—2' 2$\frac{1}{2}$''

RUN OF RAFTER _ins._	$\frac{1}{2}$	1	2	3	4	5	6	7	8	9	10	11
LENGTH OF RAFTER	$\frac{1}{2}$	$1\frac{1}{8}$	$2\frac{1}{4}$	$3\frac{3}{8}$	$4\frac{3}{8}$	$5\frac{1}{2}$	$6\frac{5}{8}$	$7\frac{3}{4}$	$8\frac{7}{8}$	$9\frac{7}{8}$	11	$12\frac{1}{8}$
LENGTH OF HIP	$\frac{3}{4}$	$1\frac{1}{2}$	3	$4\frac{1}{2}$	6	$7\frac{1}{2}$	$8\frac{7}{8}$	$10\frac{3}{8}$	$11\frac{7}{8}$	$13\frac{3}{8}$	$14\frac{7}{8}$	$16\frac{3}{8}$

RUN OF RAFTER _ft._	1	2	3	4	5	6	7	8	9	10
LENGTH OF RAFTER	$1\cdot1\frac{1}{4}$	$2\cdot2\frac{1}{2}$	$3\cdot3\frac{3}{4}$	$4\cdot5$	$5\cdot6\frac{1}{4}$	$6\cdot7\frac{1}{2}$	$7\cdot8\frac{5}{8}$	$8\cdot9\frac{7}{8}$	$9\cdot11\frac{1}{8}$	$11\cdot0\frac{3}{8}$
LENGTH OF HIP	$1\cdot5\frac{7}{8}$	$2\cdot11\frac{3}{4}$	$4\cdot5\frac{5}{8}$	$5\cdot11\frac{1}{2}$	$7\cdot5\frac{3}{8}$	$8\cdot11\frac{1}{4}$	$10\cdot5\frac{1}{8}$	$11\cdot11$	$13\cdot4\frac{3}{4}$	$14\cdot10\frac{5}{8}$

Note: Quarter Pitch (Rise = $\frac{1}{4}$ Span) has a rafter seat bevel of 26° 34′. Lengths of rafters, etc., are based on this angle

QUARTER PITCH

RISE OF COMMON RAFTER 6″ PER FOOT OF RUN

BEVELS:
1. COMMON RAFTER – SEAT $26\frac{1}{2}$
2.   ″   ″ – RIDGE $63\frac{1}{2}$
3. HIP OR VALLEY – SEAT $19\frac{1}{2}$
4.   ″ ″   ″ – RIDGE $70\frac{1}{2}$
5. JACK RAFTER – EDGE 42
6. PURLIN – EDGE 48
7.   ″ – SIDE 66

JACK RAFTERS 16 in. CENTRES DECREASE $17\frac{7}{8}$″, 18 in.—$20\frac{1}{8}$″, 24 in.—2′ $2\frac{7}{8}$″

RUN OF RAFTER	ins.	$\frac{1}{2}$	1	2	3	4	5	6	7	8	9	10	11
LENGTH OF RAFTER		$\frac{1}{2}$	$1\frac{1}{8}$	$2\frac{1}{4}$	$3\frac{3}{8}$	$4\frac{1}{2}$	$5\frac{5}{8}$	$6\frac{3}{4}$	$7\frac{7}{8}$	9	10	$11\frac{1}{4}$	$12\frac{1}{4}$
LENGTH OF HIP		$\frac{3}{4}$	$1\frac{1}{2}$	3	$4\frac{1}{2}$	6	$7\frac{1}{2}$	9	$10\frac{1}{2}$	12	$13\frac{1}{2}$	15	$16\frac{1}{2}$

RUN OF RAFTER ft.	1	2	3	4	5	6	7	8	9	10
LENGTH OF RAFTER	$1\cdot1\frac{3}{8}$	$2\cdot2\frac{7}{8}$	$3\cdot4\frac{1}{4}$	$4\cdot5\frac{5}{8}$	$5\cdot7\frac{1}{8}$	$6\cdot8\frac{1}{2}$	$7\cdot9\frac{7}{8}$	$8\cdot11\frac{3}{8}$	$10\cdot0\frac{3}{4}$	$11\cdot2\frac{1}{8}$
LENGTH OF HIP	1·6	3·0	4·6	6·0	7·6	9·0	10·6	12·0	13·6	15·0

$27\frac{1}{2}°$ PITCH

RISE OF COMMON RAFTER $6\frac{1}{4}''$ PER FOOT OF RUN

BEVELS:
1. COMMON RAFTER   – SEAT $27\frac{1}{2}$
2.   "   "   – RIDGE $62\frac{1}{2}$
3. HIP OR VALLEY   – SEAT   20
4.   "   "   "   – RIDGE   70
5. JACK RAFTER   – EDGE $41\frac{3}{4}$
6. PURLIN   – EDGE $48\frac{1}{2}$
7.   "   – SIDE   65

JACK RAFTERS 16 in. CENTRES DECREASE $18\frac{1}{8}''$, 18 in.—$20\frac{3}{8}''$, 24 in.—2′ 3″

RUN OF RAFTER	ins.	$\frac{1}{2}$	1	2	3	4	5	6	7	8	9	10	11
LENGTH OF RAFTER		$\frac{5}{8}$	$1\frac{1}{8}$	$2\frac{1}{4}$	$3\frac{3}{8}$	$4\frac{1}{2}$	$5\frac{5}{8}$	$6\frac{3}{4}$	$7\frac{7}{8}$	9	$10\frac{1}{8}$	$11\frac{1}{4}$	$12\frac{3}{8}$
LENGTH OF HIP		$\frac{3}{4}$	$1\frac{1}{2}$	3	$4\frac{1}{2}$	6	$7\frac{1}{2}$	$9\frac{1}{8}$	$10\frac{5}{8}$	12	$13\frac{1}{2}$	15	$16\frac{1}{2}$

RUN OF RAFTER	ft.	1	2	3	4	5	6	7	8	9	10
LENGTH OF RAFTER		$1·1\frac{1}{2}$	$2·3$	$3·4\frac{1}{2}$	$4·6\frac{1}{8}$	$5·7\frac{5}{8}$	$6·9\frac{1}{8}$	$7·10\frac{5}{8}$	$9·0\frac{1}{4}$	$10·1\frac{3}{4}$	$11·3\frac{1}{4}$
LENGTH OF HIP		$1·6$	$3·0$	$4·6\frac{1}{8}$	$6·0\frac{3}{8}$	$7·6\frac{3}{8}$	$9·0\frac{1}{2}$	$10·6\frac{5}{8}$	$12·0\frac{5}{8}$	$13·6\frac{3}{4}$	$15·0\frac{7}{8}$

28° PITCH

RISE OF COMMON RAFTER $6\frac{3}{8}''$ PER FOOT OF RUN

BEVELS:
1. COMMON RAFTER   – SEAT   28
2.   "   "   – RIDGE   62
3. HIP OR VALLEY   – SEAT $20\frac{1}{2}$
4.   "   "   "   – RIDGE $69\frac{1}{2}$
5. JACK RAFTER   – EDGE $41\frac{1}{2}$
6. PURLIN   – EDGE $48\frac{1}{2}$
7.   "   – SIDE   65

JACK RAFTERS 16 in. CENTRES DECREASE $18\frac{1}{8}''$, 18 in.—$20\frac{3}{8}''$, 24 in.—$2' \, 3\frac{1}{8}''$

RUN OF RAFTER	ins.	$\frac{1}{2}$	1	2	3	4	5	6	7	8	9	10	11
LENGTH OF RAFTER		$\frac{5}{8}$	$1\frac{1}{8}$	$2\frac{1}{4}$	$3\frac{3}{8}$	$4\frac{1}{2}$	$5\frac{5}{8}$	$6\frac{3}{4}$	$7\frac{7}{8}$	9	$10\frac{1}{4}$	$11\frac{3}{8}$	$12\frac{1}{2}$
LENGTH OF HIP		$\frac{3}{4}$	$1\frac{1}{2}$	3	$4\frac{1}{2}$	6	$7\frac{1}{2}$	$9\frac{1}{8}$	$10\frac{5}{8}$	$12\frac{1}{8}$	$13\frac{5}{8}$	$15\frac{1}{8}$	$16\frac{5}{8}$

RUN OF RAFTER	ft.	1	2	3	4	5	6	7	8	9	10
LENGTH OF RAFTER		$1 \cdot 1\frac{5}{8}$	$2 \cdot 3\frac{1}{8}$	$3 \cdot 4\frac{3}{4}$	$4 \cdot 6\frac{3}{8}$	$5 \cdot 8$	$6 \cdot 9\frac{1}{2}$	$7 \cdot 11\frac{1}{8}$	$9 \cdot 0\frac{3}{4}$	$10 \cdot 2\frac{3}{8}$	$11 \cdot 3\frac{7}{8}$
LENGTH OF HIP		$1 \cdot 6\frac{1}{8}$	$3 \cdot 0\frac{1}{4}$	$4 \cdot 6\frac{3}{8}$	$6 \cdot 0\frac{1}{2}$	$7 \cdot 6\frac{5}{8}$	$9 \cdot 0\frac{3}{4}$	$10 \cdot 6\frac{7}{8}$	$12 \cdot 1$	$13 \cdot 7\frac{1}{8}$	$15 \cdot 1\frac{3}{8}$

29° PITCH

RISE OF COMMON RAFTER 6⅝″ PER FOOT OF RUN

BEVELS :
1. COMMON RAFTER – SEAT 29
2.   ″   ″ – RIDGE 61
3. HIP OR VALLEY – SEAT $21\frac{1}{2}$
4. ″   ″   ″ – RIDGE $68\frac{1}{2}$
5. JACK RAFTER – EDGE 41
6. PURLIN – EDGE 49
7.   ″ – SIDE 64

JACK RAFTERS 16 in. CENTRES DECREASE $18\frac{1}{4}$″, 18 in.—$20\frac{5}{8}$″, 24 in.—2′ $3\frac{1}{2}$″

RUN OF RAFTER	ins.	½	1	2	3	4	5	6	7	8	9	10	11
LENGTH OF RAFTER		$\frac{5}{8}$	$1\frac{1}{8}$	$2\frac{1}{4}$	$3\frac{3}{8}$	$4\frac{5}{8}$	$5\frac{3}{4}$	$6\frac{7}{8}$	8	$9\frac{1}{8}$	$10\frac{1}{4}$	$11\frac{1}{2}$	$12\frac{5}{8}$
LENGTH OF HIP		$\frac{3}{4}$	$1\frac{1}{2}$	3	$4\frac{5}{8}$	$6\frac{1}{8}$	$7\frac{5}{8}$	$9\frac{1}{8}$	$10\frac{5}{8}$	$12\frac{1}{8}$	$13\frac{5}{8}$	$15\frac{1}{4}$	$16\frac{3}{4}$

RUN OF RAFTER	ft.	1	2	3	4	5	6	7	8	9	10
LENGTH OF RAFTER		$1.1\frac{3}{4}$	$2.3\frac{1}{2}$	$3.5\frac{1}{8}$	$4.6\frac{7}{8}$	$5.8\frac{5}{8}$	$6.10\frac{3}{8}$	8.0	$9.1\frac{3}{4}$	$10.3\frac{1}{2}$	$11.5\frac{1}{4}$
LENGTH OF HIP		$1.6\frac{1}{4}$	$3.0\frac{1}{2}$	$4.6\frac{5}{8}$	$6.0\frac{7}{8}$	$7.7\frac{1}{8}$	$9.1\frac{3}{8}$	$10.7\frac{5}{8}$	$12.1\frac{7}{8}$	13.8	$15.2\frac{1}{4}$

30° PITCH

RISE OF COMMON RAFTER $6\frac{15}{16}''$ PER FOOT OF RUN

BEVELS:
1. COMMON RAFTER — SEAT 30
2.   ″   ″ — RIDGE 60
3. HIP OR VALLEY — SEAT 22
4.   ″   ″   ″ — RIDGE 68
5. JACK RAFTER — EDGE 41
6. PURLIN — EDGE 49
7.   ″ — SIDE $63\frac{1}{2}$

JACK RAFTERS 16 in. CENTRES DECREASE $18\frac{1}{2}''$, 18 in.—$20\frac{3}{4}''$, 24 in.—2′ $3\frac{3}{4}''$

RUN OF RAFTER	*ins.*	$\frac{1}{2}$	1	2	3	4	5	6	7	8	9	10	11
LENGTH OF RAFTER … …		$\frac{5}{8}$	$1\frac{1}{8}$	$2\frac{1}{4}$	$3\frac{1}{2}$	$4\frac{5}{8}$	$5\frac{3}{4}$	$6\frac{7}{8}$	$8\frac{1}{8}$	$9\frac{1}{4}$	$10\frac{3}{8}$	$11\frac{1}{2}$	$12\frac{3}{4}$
LENGTH OF HIP … …		$\frac{3}{4}$	$1\frac{1}{2}$	3	$4\frac{5}{8}$	$6\frac{1}{8}$	$7\frac{5}{8}$	$9\frac{1}{8}$	$10\frac{3}{4}$	$12\frac{1}{4}$	$13\frac{3}{4}$	$15\frac{1}{4}$	$16\frac{3}{4}$

RUN OF RAFTER	*ft.*	1	2	3	4	5	6	7	8	9	10
LENGTH OF RAFTER		$1 \cdot 1\frac{7}{8}$	$2 \cdot 3\frac{3}{4}$	$3 \cdot 5\frac{5}{8}$	$4 \cdot 7\frac{3}{8}$	$5 \cdot 9\frac{1}{4}$	$6 \cdot 11\frac{1}{8}$	$8 \cdot 1$	$9 \cdot 2\frac{7}{8}$	$10 \cdot 4\frac{3}{4}$	$11 \cdot 6\frac{5}{8}$
LENGTH OF HIP		$1 \cdot 6\frac{3}{8}$	$3 \cdot 0\frac{5}{8}$	$4 \cdot 7$	$6 \cdot 1\frac{1}{4}$	$7 \cdot 7\frac{5}{8}$	$9 \cdot 2$	$10 \cdot 8\frac{3}{8}$	$12 \cdot 2\frac{5}{8}$	$13 \cdot 9$	$15 \cdot 3\frac{1}{4}$

31° PITCH

RISE OF COMMON RAFTER $7\frac{3}{16}''$ PER FOOT OF RUN

BEVELS :
1. COMMON RAFTER   – SEAT   31
2. $''$ $''$ – RIDGE   59
3. HIP OR VALLEY   – SEAT   23
4. $''$ $''$ $''$ – RIDGE   67
5. JACK RAFTER   – EDGE $40\frac{1}{2}$
6. PURLIN   – EDGE $49\frac{1}{2}$
7. $''$ – SIDE $62\frac{1}{2}$

JACK RAFTERS 16 in. CENTRES DECREASE $18\frac{5}{8}''$, 18 in.—21″, 24 in.—2′ 4″

RUN OF RAFTER	*ins.*	$\frac{1}{2}$	1	2	3	4	5	6	7	8	9	10	11
LENGTH OF RAFTER		$\frac{5}{8}$	$1\frac{1}{8}$	$2\frac{3}{8}$	$3\frac{1}{2}$	$4\frac{5}{8}$	$5\frac{7}{8}$	7	$8\frac{1}{8}$	$9\frac{3}{8}$	$10\frac{1}{2}$	$11\frac{5}{8}$	$12\frac{7}{8}$
LENGTH OF HIP		$\frac{3}{4}$	$1\frac{1}{2}$	$3\frac{1}{8}$	$4\frac{5}{8}$	$6\frac{1}{8}$	$7\frac{5}{8}$	$9\frac{1}{4}$	$10\frac{3}{4}$	$12\frac{1}{4}$	$13\frac{7}{8}$	$15\frac{3}{8}$	$16\frac{7}{8}$

RUN OF RAFTER	*ft.*	1	2	3	4	5	6	7	8	9	10
LENGTH OF RAFTER		1·2	2·4	3·6	4·8	5·10	7·0	8·2	9·4	10·6	11·8
LENGTH OF HIP		$1·6\frac{1}{2}$	$3·0\frac{7}{8}$	$4·7\frac{1}{4}$	$6·1\frac{3}{4}$	$7·8\frac{1}{8}$	$9·2\frac{5}{8}$	10·9	$12·3\frac{1}{2}$	13·10	$15·4\frac{3}{8}$

32° PITCH

RISE OF COMMON RAFTER 7½″ PER FOOT OF RUN

BEVELS:
1. COMMON RAFTER – SEAT 32
2.   ″   ″ – RIDGE 58
3. HIP OR VALLEY – SEAT 24
4.   ″   ″   ″ – RIDGE 66
5. JACK RAFTER – EDGE 40½
6. PURLIN – EDGE 49½
7.   ″ – SIDE 62

JACK RAFTERS 16 in. CENTRES DECREASE 18⅞″, 18 in.—21¼″, 24 in.—2′ 4¼″

RUN OF RAFTER	*ins.*	½	1	2	3	4	5	6	7	8	9	10	11
LENGTH OF RAFTER		⅝	1⅛	2⅜	3½	4¾	5⅞	7⅛	8¼	9⅜	10⅝	11¾	13
LENGTH OF HIP		¾	1½	3⅛	4⅝	6⅛	7¾	9¼	10⅞	12⅜	13⅞	15½	17

RUN OF RAFTER	*ft.*	1	2	3	4	5	6	7	8	9	10
LENGTH OF RAFTER		1·2⅛	2·4¼	3·6½	4·8⅝	5·10¾	7·0⅞	8·3	9·5¼	10·7⅜	11·9½
LENGTH OF HIP		1·6½	3·1⅛	4·7⅝	6·2¼	7·8¾	9·3⅜	10·9⅞	12·4½	13·11	15·5½

$32\frac{1}{2}^\circ$ PITCH

RISE OF COMMON RAFTER $7\frac{5}{8}''$ PER FOOT OF RUN

BEVELS:
1. COMMON RAFTER   – SEAT $32\frac{1}{2}$
2.   "   "   – RIDGE $57\frac{1}{2}$
3. HIP OR VALLEY   – SEAT $24\frac{1}{2}$
4.   "   "   "   – RIDGE $65\frac{3}{4}$
5. JACK RAFTER   – EDGE $40\frac{1}{4}$
6. PURLIN   – EDGE $49\frac{3}{4}$
7.   "   – SIDE $61\frac{1}{2}$

JACK RAFTERS 16 in. CENTRES DECREASE 19″, 18 in.—$21\frac{3}{8}''$, 24 in.—2′ $4\frac{1}{2}''$

RUN OF RAFTER	ins.	$\frac{1}{2}$	1	2	3	4	5	6	7	8	9	10	11
LENGTH OF RAFTER		$\frac{5}{8}$	$1\frac{1}{8}$	$2\frac{3}{8}$	$3\frac{1}{2}$	$4\frac{3}{4}$	$5\frac{7}{8}$	$7\frac{1}{8}$	$8\frac{1}{4}$	$9\frac{1}{2}$	$10\frac{5}{8}$	$11\frac{7}{8}$	13
LENGTH OF HIP		$\frac{3}{4}$	$1\frac{1}{2}$	$3\frac{1}{8}$	$4\frac{5}{8}$	$6\frac{1}{8}$	$7\frac{3}{4}$	$9\frac{1}{4}$	$10\frac{7}{8}$	$12\frac{3}{8}$	14	$15\frac{1}{2}$	17

RUN OF RAFTER	ft.	1	2	3	4	5	6	7	8	9	10
LENGTH OF RAFTER		$1.2\frac{1}{4}$	$2.4\frac{1}{2}$	$3.6\frac{5}{8}$	$4.8\frac{7}{8}$	$5.11\frac{1}{8}$	$7.1\frac{3}{8}$	$8.3\frac{5}{8}$	$9.5\frac{7}{8}$	10.8	$11.10\frac{1}{4}$
LENGTH OF HIP		$1.6\frac{5}{8}$	$3.1\frac{1}{4}$	$4.7\frac{7}{8}$	$6.2\frac{3}{8}$	7.9	$9.3\frac{5}{8}$	$10.10\frac{1}{4}$	$12.4\frac{7}{8}$	$13.11\frac{1}{2}$	$15.6\frac{1}{8}$

33° PITCH

RISE OF COMMON RAFTER $7\frac{13}{16}''$ PER FOOT OF RUN

BEVELS:	1.	COMMON RAFTER	– SEAT	33
	2.	" "	– RIDGE	57
	3.	HIP OR VALLEY	– SEAT	$24\frac{1}{2}$
	4.	" " "	– RIDGE	$65\frac{1}{2}$
	5.	JACK RAFTER	– EDGE	40
	6.	PURLIN	– EDGE	50
	7.	"	– SIDE	$61\frac{1}{2}$

JACK RAFTERS 16 in. CENTRES DECREASE $19\frac{1}{8}''$, 18 in.—$21\frac{1}{2}''$, 24 in.—$2'\ 4\frac{5}{8}''$

RUN OF RAFTER	ins.	$\frac{1}{2}$	1	2	3	4	5	6	7	8	9	10	11
LENGTH OF RAFTER		$\frac{5}{8}$	$1\frac{1}{4}$	$2\frac{3}{8}$	$3\frac{5}{8}$	$4\frac{3}{4}$	6	$7\frac{1}{8}$	$8\frac{3}{8}$	$9\frac{1}{2}$	$10\frac{3}{4}$	$11\frac{7}{8}$	$13\frac{1}{8}$
LENGTH OF HIP		$\frac{3}{4}$	$1\frac{1}{2}$	$3\frac{1}{8}$	$4\frac{5}{8}$	$6\frac{1}{4}$	$7\frac{3}{4}$	$9\frac{3}{8}$	$10\frac{7}{8}$	$12\frac{1}{2}$	14	$15\frac{1}{2}$	$17\frac{1}{8}$

RUN OF RAFTER	ft.	1	2	3	4	5	6	7	8	9	10
LENGTH OF RAFTER		$1.2\frac{1}{4}$	$2.4\frac{5}{8}$	$3.6\frac{7}{8}$	$4.9\frac{1}{4}$	$5.11\frac{1}{2}$	$7.1\frac{7}{8}$	$8.4\frac{1}{8}$	$9.6\frac{1}{2}$	$10.8\frac{3}{4}$	$11.11\frac{1}{8}$
LENGTH OF HIP		$1.6\frac{5}{8}$	$3.1\frac{3}{8}$	4.8	$6.2\frac{3}{8}$	$7.9\frac{3}{8}$	9.4	$10.10\frac{5}{8}$	$12.5\frac{3}{8}$	14.0	$15.6\frac{5}{8}$

Note: Third Pitch (Rise = $\frac{1}{3}$ Span) has a rafter seat bevel of 33° 40′. Lengths of rafters, etc., are based on this angle

THIRD PITCH

RISE OF COMMON RAFTER 8″ PER FOOT OF RUN

BEVELS:
1. COMMON RAFTER — SEAT $33\frac{1}{2}$
2.   ″   ″ — RIDGE $56\frac{1}{2}$
3. HIP OR VALLEY — SEAT 25
4. ″ ″ ″ — RIDGE 65
5. JACK RAFTER — EDGE 40
6. PURLIN — EDGE 50
7.   ″ — SIDE 61

JACK RAFTERS 16 in. CENTRES DECREASE $19\frac{1}{4}$″, 18 in.—$21\frac{5}{8}$″, 24 in.—2′ $4\frac{7}{8}$″

RUN OF RAFTER ins.	$\frac{1}{2}$	1	2	3	4	5	6	7	8	9	10	11
LENGTH OF RAFTER … …	$\frac{5}{8}$	$1\frac{1}{4}$	$2\frac{3}{8}$	$3\frac{5}{8}$	$4\frac{3}{4}$	6	$7\frac{1}{4}$	$8\frac{3}{8}$	$9\frac{5}{8}$	$10\frac{7}{8}$	12	$13\frac{1}{4}$
LENGTH OF HIP … …	$\frac{3}{4}$	$1\frac{1}{2}$	$3\frac{1}{8}$	$4\frac{5}{8}$	$6\frac{1}{4}$	$7\frac{7}{8}$	$9\frac{3}{8}$	$10\frac{7}{8}$	$12\frac{1}{2}$	$14\frac{1}{8}$	$15\frac{5}{8}$	$17\frac{1}{4}$

RUN OF RAFTER ft.	1	2	3	4	5	6	7	8	9	10
LENGTH OF RAFTER	$1·2\frac{3}{8}$	$2·4\frac{7}{8}$	$3·7\frac{1}{4}$	$4·9\frac{5}{8}$	$6·0\frac{1}{8}$	$7·2\frac{1}{2}$	$8·4\frac{7}{8}$	$9·7\frac{3}{8}$	$10·9\frac{3}{4}$	$12·0\frac{1}{8}$
LENGTH OF HIP	$1·6\frac{3}{4}$	$3·1\frac{1}{2}$	$4·8\frac{1}{4}$	6·3	$7·9\frac{3}{4}$	$9·4\frac{1}{2}$	$10·11\frac{1}{4}$	12·6	$14·0\frac{7}{8}$	$15·7\frac{5}{8}$

35° PITCH

RISE OF COMMON RAFTER 8$\frac{7}{16}$″ PER FOOT OF RUN

BEVELS:
1. COMMON RAFTER – SEAT 35
2.   ″   ″ – RIDGE 55
3. HIP OR VALLEY – SEAT 26$\frac{1}{2}$
4. ″   ″   ″ – RIDGE 63$\frac{1}{2}$
5. JACK RAFTER – EDGE 39$\frac{1}{2}$
6. PURLIN – EDGE 50$\frac{1}{2}$
7.   ″ – SIDE 60

JACK RAFTERS 16 in. CENTRES DECREASE 19$\frac{1}{2}$″, 18 in.—22″, 24 in.—2′ 5$\frac{1}{4}$″

RUN OF RAFTER	ins.	$\frac{1}{2}$	1	2	3	4	5	6	7	8	9	10	11
LENGTH OF RAFTER		$\frac{5}{8}$	1$\frac{1}{4}$	2$\frac{1}{2}$	3$\frac{5}{8}$	4$\frac{7}{8}$	6$\frac{1}{8}$	7$\frac{3}{8}$	8$\frac{1}{2}$	9$\frac{3}{4}$	11	12$\frac{1}{4}$	13$\frac{3}{8}$
LENGTH OF HIP		$\frac{3}{4}$	1$\frac{5}{8}$	3$\frac{1}{8}$	4$\frac{3}{4}$	6$\frac{3}{8}$	7$\frac{7}{8}$	9$\frac{1}{2}$	11	12$\frac{5}{8}$	14$\frac{1}{4}$	15$\frac{3}{4}$	17$\frac{3}{8}$

RUN OF RAFTER	ft.	1	2	3	4	5	6	7	8	9	10
LENGTH OF RAFTER		1·2$\frac{5}{8}$	2·5$\frac{1}{4}$	3·8	4·10$\frac{5}{8}$	6·1$\frac{1}{4}$	7·3$\frac{7}{8}$	8·6$\frac{1}{2}$	9·9$\frac{1}{4}$	10·11$\frac{7}{8}$	12·2$\frac{1}{2}$
LENGTH OF HIP		1·6$\frac{7}{8}$	3·1$\frac{7}{8}$	4·8$\frac{3}{4}$	6·3$\frac{3}{4}$	7·10$\frac{5}{8}$	9·5$\frac{5}{8}$	11·0$\frac{1}{2}$	12·7$\frac{1}{2}$	14·2$\frac{1}{2}$	15·9$\frac{3}{8}$

36° PITCH

RISE OF COMMON RAFTER $8\frac{3}{4}''$ PER FOOT OF RUN

BEVELS:	1.	COMMON RAFTER	– SEAT	36
	2.	" "	– RIDGE	54
	3.	HIP OR VALLEY	– SEAT	27
	4.	" " "	– RIDGE	63
	5.	JACK RAFTER	– EDGE	39
	6.	PURLIN	– EDGE	51
	7.	"	– SIDE	$59\frac{1}{2}$

JACK RAFTERS 16 in. CENTRES DECREASE $19\frac{3}{4}''$, 18 in.—$22\frac{1}{4}''$, 24 in.—$2' 5\frac{5}{8}''$

RUN OF RAFTER	ins.	$\frac{1}{2}$	1	2	3	4	5	6	7	8	9	10	11
LENGTH OF RAFTER		$\frac{5}{8}$	$1\frac{1}{4}$	$2\frac{1}{2}$	$3\frac{3}{4}$	5	$6\frac{1}{8}$	$7\frac{3}{8}$	$8\frac{5}{8}$	$9\frac{7}{8}$	$11\frac{1}{8}$	$12\frac{3}{8}$	$13\frac{5}{8}$
LENGTH OF HIP		$\frac{3}{4}$	$1\frac{5}{8}$	$3\frac{1}{8}$	$4\frac{3}{4}$	$6\frac{3}{8}$	8	$9\frac{1}{2}$	$11\frac{1}{8}$	$12\frac{3}{4}$	$14\frac{1}{4}$	$15\frac{7}{8}$	$17\frac{1}{2}$

RUN OF RAFTER	ft.	1	2	3	4	5	6	7	8	9	10
LENGTH OF RAFTER		$1.2\frac{7}{8}$	$2.5\frac{5}{8}$	$3.8\frac{1}{2}$	$4.11\frac{3}{8}$	$6.2\frac{1}{8}$	7.5	$8.7\frac{7}{8}$	$9.10\frac{5}{8}$	$11.1\frac{1}{2}$	$12.4\frac{3}{8}$
LENGTH OF HIP		$1.7\frac{1}{8}$	$3.2\frac{1}{8}$	$4.9\frac{1}{4}$	$6.4\frac{1}{4}$	$7.11\frac{3}{8}$	$9.6\frac{1}{2}$	$11.1\frac{1}{2}$	$12.8\frac{5}{8}$	$14.3\frac{5}{8}$	$15.10\frac{3}{4}$

37° PITCH

RISE OF COMMON RAFTER 9$\frac{1}{16}$″ PER FOOT OF RUN

BEVELS:
1. COMMON RAFTER – SEAT 37
2.   ″   ″   – RIDGE 53
3. HIP OR VALLEY – SEAT 28
4.   ″   ″   ″   – RIDGE 62
5. JACK RAFTER – EDGE 38$\frac{1}{2}$
6. PURLIN – EDGE 51$\frac{1}{2}$
7.   ″   – SIDE 59

JACK RAFTERS 16 in. CENTRES DECREASE 20″, 18 in.—22$\frac{1}{2}$″, 24 in.—2′ 6″

RUN OF RAFTER	ins.	$\frac{1}{2}$	1	2	3	4	5	6	7	8	9	10	11
LENGTH OF RAFTER		$\frac{5}{8}$	1$\frac{1}{4}$	2$\frac{1}{2}$	3$\frac{3}{4}$	5	6$\frac{1}{4}$	7$\frac{1}{2}$	8$\frac{3}{4}$	10	11$\frac{1}{4}$	12$\frac{1}{2}$	13$\frac{3}{4}$
LENGTH OF HIP		$\frac{3}{4}$	1$\frac{5}{8}$	3$\frac{1}{4}$	4$\frac{3}{4}$	6$\frac{3}{8}$	8	9$\frac{5}{8}$	11$\frac{1}{4}$	12$\frac{7}{8}$	14$\frac{3}{8}$	16	17$\frac{5}{8}$

RUN OF RAFTER	ft.	1	2	3	4	5	6	7	8	9	10
LENGTH OF RAFTER		1·3	2·6	3·9$\frac{1}{8}$	5·0$\frac{1}{8}$	6·3$\frac{1}{8}$	7·6$\frac{1}{8}$	8·9$\frac{1}{8}$	10·0$\frac{1}{4}$	11·3$\frac{1}{4}$	12·6$\frac{1}{4}$
LENGTH OF HIP		1·7$\frac{1}{4}$	3·2$\frac{1}{2}$	4·9$\frac{5}{8}$	6·4$\frac{7}{8}$	8·0$\frac{1}{8}$	9·7$\frac{3}{8}$	11·2$\frac{5}{8}$	12·9$\frac{7}{8}$	14·5	16·0$\frac{1}{4}$

$37\frac{1}{2}°$ PITCH

RISE OF COMMON RAFTER $9\frac{3}{16}''$ PER FOOT OF RUN

BEVELS:
1. COMMON RAFTER – SEAT $37\frac{1}{2}$
2.   "   " – RIDGE $52\frac{1}{2}$
3. HIP OR VALLEY – SEAT $28\frac{1}{2}$
4.   " " " – RIDGE $61\frac{1}{2}$
5. JACK RAFTER – EDGE $38\frac{1}{4}$
6. PURLIN – EDGE $51\frac{3}{4}$
7.   " – SIDE $58\frac{1}{2}$

JACK RAFTERS 16 in. CENTRES DECREASE $20\frac{1}{8}''$, 18 in.—$22\frac{3}{4}''$, 24 in.—$2'\,6\frac{1}{4}''$

RUN OF RAFTER ins.	$\frac{1}{2}$	1	2	3	4	5	6	7	8	9	10	11
LENGTH OF RAFTER	$\frac{5}{8}$	$1\frac{1}{4}$	$2\frac{1}{2}$	$3\frac{3}{4}$	5	$6\frac{1}{4}$	$7\frac{1}{2}$	$8\frac{3}{4}$	10	$11\frac{3}{8}$	$12\frac{5}{8}$	14
LENGTH OF HIP	$\frac{3}{4}$	$1\frac{5}{8}$	$3\frac{1}{4}$	$4\frac{3}{4}$	$6\frac{3}{8}$	8	$9\frac{5}{8}$	$11\frac{1}{4}$	$12\frac{7}{8}$	$14\frac{1}{2}$	$16\frac{1}{8}$	$17\frac{3}{4}$

RUN OF RAFTER ft.	1	2	3	4	5	6	7	8	9	10
LENGTH OF RAFTER	$1.3\frac{1}{8}$	$2.6\frac{1}{4}$	$3.9\frac{3}{8}$	$5.0\frac{1}{2}$	$6.3\frac{5}{8}$	$7.6\frac{3}{4}$	$8.9\frac{7}{8}$	10.1	$11.4\frac{1}{8}$	$12.7\frac{1}{4}$
LENGTH OF HIP	$1.7\frac{3}{8}$	$3.2\frac{5}{8}$	4.10	$6.5\frac{1}{4}$	$8.0\frac{1}{2}$	$9.7\frac{7}{8}$	$11.3\frac{1}{4}$	$12.10\frac{1}{2}$	$14.5\frac{3}{4}$	$16.1\frac{1}{8}$

38° PITCH

RISE OF COMMON RAFTER $9\frac{3}{8}''$ PER FOOT OF RUN

BEVELS:
1. COMMON RAFTER   – SEAT   38
2.   "   "   – RIDGE   52
3. HIP OR VALLEY   – SEAT   29
4.   "   "   "   – RIDGE   61
5. JACK RAFTER   – EDGE   38
6. PURLIN   – EDGE   51
7.   "   – SIDE $58\frac{1}{2}$

JACK RAFTERS 16 in. CENTRES DECREASE $20\frac{1}{4}''$, 18 in.—$22\frac{7}{8}''$, 24 in.—$2'\,6\frac{1}{2}''$

RUN OF RAFTER	ins.	$\frac{1}{2}$	1	2	3	4	5	6	7	8	9	10	11
LENGTH OF RAFTER		$\frac{5}{8}$	$1\frac{1}{4}$	$2\frac{1}{2}$	$3\frac{3}{4}$	$5\frac{1}{8}$	$6\frac{3}{8}$	$7\frac{5}{8}$	$8\frac{7}{8}$	$10\frac{1}{8}$	$11\frac{3}{8}$	$12\frac{3}{4}$	14
LENGTH OF HIP		$\frac{3}{4}$	$1\frac{5}{8}$	$3\frac{1}{4}$	$4\frac{7}{8}$	$6\frac{1}{2}$	$8\frac{1}{8}$	$9\frac{3}{4}$	$11\frac{1}{4}$	$12\frac{7}{8}$	$14\frac{1}{2}$	$16\frac{1}{8}$	$17\frac{3}{4}$

RUN OF RAFTER	ft.	1	2	3	4	5	6	7	8	9	10
LENGTH OF RAFTER		$1 \cdot 3\frac{1}{4}$	$2 \cdot 6\frac{1}{2}$	$3 \cdot 9\frac{5}{8}$	$5 \cdot 0\frac{7}{8}$	$6 \cdot 4\frac{1}{8}$	$7 \cdot 7\frac{3}{8}$	$8 \cdot 10\frac{5}{8}$	$10 \cdot 1\frac{7}{8}$	$11 \cdot 5$	$12 \cdot 8\frac{1}{4}$
LENGTH OF HIP		$1 \cdot 7\frac{3}{8}$	$3 \cdot 2\frac{3}{4}$	$4 \cdot 10\frac{1}{8}$	$6 \cdot 5\frac{1}{2}$	$8 \cdot 0\frac{7}{8}$	$9 \cdot 8\frac{1}{4}$	$11 \cdot 3\frac{5}{8}$	$12 \cdot 11$	$14 \cdot 6\frac{3}{8}$	$16 \cdot 1\frac{3}{4}$

39° PITCH

RISE OF COMMON RAFTER $9\frac{3}{4}''$ PER FOOT OF RUN

BEVELS:
1. COMMON RAFTER – SEAT 39
2.   ''   ''   – RIDGE 51
3. HIP OR VALLEY – SEAT 30
4. ''   ''   '' – RIDGE 60
5. JACK RAFTER – EDGE 38
6. PURLIN – EDGE 52
7.   ''   – SIDE 58

JACK RAFTERS 16 in. CENTRES DECREASE $20\frac{5}{8}''$, 18 in.—$23\frac{1}{8}''$, 24 in.—$2'\,6\frac{7}{8}''$

RUN OF RAFTER	ins.	$\frac{1}{2}$	1	2	3	4	5	6	7	8	9	10	11
LENGTH OF RAFTER		$\frac{5}{8}$	$1\frac{1}{4}$	$2\frac{5}{8}$	$3\frac{7}{8}$	$5\frac{1}{8}$	$6\frac{3}{8}$	$7\frac{3}{4}$	9	$10\frac{1}{4}$	$11\frac{5}{8}$	$12\frac{7}{8}$	$14\frac{1}{8}$
LENGTH OF HIP		$\frac{3}{4}$	$1\frac{5}{8}$	$3\frac{1}{4}$	$4\frac{7}{8}$	$6\frac{1}{2}$	$8\frac{1}{8}$	$9\frac{3}{4}$	$11\frac{3}{8}$	13	$14\frac{5}{8}$	$16\frac{1}{4}$	$17\frac{7}{8}$

RUN OF RAFTER	ft.	1	2	3	4	5	6	7	8	9	10
LENGTH OF RAFTER		$1\cdot3\frac{1}{2}$	$2\cdot6\frac{7}{8}$	$3\cdot10\frac{3}{8}$	$5\cdot1\frac{3}{4}$	$6\cdot5\frac{1}{4}$	$7\cdot8\frac{5}{8}$	$9\cdot0\frac{1}{8}$	$10\cdot3\frac{1}{2}$	$11\cdot7$	$12\cdot10\frac{3}{8}$
LENGTH OF HIP		$1\cdot7\frac{1}{2}$	$3\cdot3\frac{1}{8}$	$4\cdot10\frac{5}{8}$	$6\cdot6\frac{1}{8}$	$8\cdot1\frac{3}{4}$	$9\cdot9\frac{1}{4}$	$11\cdot4\frac{7}{8}$	$13\cdot0\frac{3}{8}$	$14\cdot7\frac{7}{8}$	$16\cdot3\frac{1}{2}$

40° PITCH

RISE OF COMMON RAFTER $10\frac{1}{16}''$ PER FOOT OF RUN

BEVELS:
1. COMMON RAFTER   – SEAT   40
2.   ″   ″   – RIDGE   50
3. HIP OR VALLEY   – SEAT $30\frac{1}{2}$
4.   ″   ″   ″   – RIDGE $59\frac{1}{2}$
5. JACK RAFTER   – EDGE $37\frac{1}{2}$
6. PURLIN   – EDGE $52\frac{1}{2}$
7.   ″   – SIDE $57\frac{1}{2}$

JACK RAFTERS 16 in. CENTRES DECREASE $20\frac{7}{8}''$, 18 in.—$23\frac{1}{2}''$, 24 in.—$2'\,7\frac{3}{8}''$

RUN OF RAFTER *ins.*	$\frac{1}{2}$	1	2	3	4	5	6	7	8	9	10	11
LENGTH OF RAFTER	$\frac{5}{8}$	$1\frac{1}{4}$	$2\frac{5}{8}$	$3\frac{7}{8}$	$5\frac{1}{4}$	$6\frac{1}{2}$	$7\frac{7}{8}$	$9\frac{1}{8}$	$10\frac{1}{2}$	$11\frac{3}{4}$	13	$14\frac{3}{8}$
LENGTH OF HIP	$\frac{7}{8}$	$1\frac{5}{8}$	$3\frac{1}{4}$	$4\frac{7}{8}$	$6\frac{5}{8}$	$8\frac{1}{4}$	$9\frac{7}{8}$	$11\frac{1}{2}$	$13\frac{1}{8}$	$14\frac{3}{4}$	$16\frac{1}{2}$	$18\frac{1}{8}$

RUN OF RAFTER *ft.*	1	2	3	4	5	6	7	8	9	10
LENGTH OF RAFTER	$1.3\frac{5}{8}$	$2.7\frac{3}{8}$	3.11	$5.2\frac{5}{8}$	$6.6\frac{3}{8}$	7.10	$9.1\frac{5}{8}$	$10.5\frac{3}{8}$	11.9	$13.0\frac{5}{8}$
LENGTH OF HIP	$1.7\frac{3}{4}$	$3.3\frac{1}{2}$	$4.11\frac{1}{8}$	$6.6\frac{7}{8}$	$8.2\frac{5}{8}$	$9.10\frac{3}{8}$	$11.6\frac{1}{8}$	$13.1\frac{7}{8}$	$14.9\frac{1}{2}$	$16.5\frac{1}{4}$

41° PITCH

RISE OF COMMON RAFTER 10$\frac{7}{16}$″ PER FOOT OF RUN

BEVELS:
1. COMMON RAFTER – SEAT 41
2.   ″   ″ – RIDGE 49
3. HIP OR VALLEY – SEAT 31$\frac{1}{2}$
4. ″ ″   ″ – RIDGE 58$\frac{1}{2}$
5. JACK RAFTER – EDGE 37
6. PURLIN – EDGE 53
7.   ″ – SIDE 56$\frac{1}{2}$

JACK RAFTERS 16 in. CENTRES DECREASE 21$\frac{1}{4}$″, 18 in.—23$\frac{7}{8}$″, 24 in.—2′ 7$\frac{3}{4}$″

RUN OF RAFTER	*ins.*	$\frac{1}{2}$	1	2	3	4	5	6	7	8	9	10	11
LENGTH OF RAFTER		$\frac{5}{8}$	1$\frac{3}{8}$	2$\frac{5}{8}$	4	5$\frac{1}{4}$	6$\frac{5}{8}$	8	9$\frac{1}{4}$	10$\frac{5}{8}$	11$\frac{7}{8}$	13$\frac{1}{4}$	14$\frac{5}{8}$
LENGTH OF HIP		$\frac{7}{8}$	1$\frac{5}{8}$	3$\frac{3}{8}$	5	6$\frac{5}{8}$	8$\frac{1}{4}$	10	11$\frac{5}{8}$	13$\frac{1}{4}$	15	16$\frac{5}{8}$	18$\frac{1}{4}$

RUN OF RAFTER	*ft.*	1	2	3	4	5	6	7	8	9	10
LENGTH OF RAFTER		1·3$\frac{7}{8}$	2·7$\frac{3}{4}$	3·11$\frac{3}{4}$	5·3$\frac{5}{8}$	6·7$\frac{1}{2}$	7·11$\frac{3}{8}$	9·3$\frac{1}{4}$	10·7$\frac{1}{4}$	11·11$\frac{1}{8}$	13·3
LENGTH OF HIP		1·7$\frac{7}{8}$	3·3$\frac{7}{8}$	4·11$\frac{3}{4}$	6·7$\frac{5}{8}$	8·3$\frac{5}{8}$	9·11$\frac{1}{2}$	11·7$\frac{1}{2}$	13·3$\frac{3}{8}$	14·11$\frac{1}{4}$	16·7$\frac{1}{4}$

42° PITCH

RISE OF COMMON RAFTER 10$\frac{13}{16}''$ PER FOOT OF RUN

BEVELS :
1. COMMON RAFTER – SEAT 42
2.   "   " – RIDGE 48
3. HIP OR VALLEY – SEAT 32$\frac{1}{2}$
4.   "   "   " – RIDGE 57$\frac{1}{2}$
5. JACK RAFTER – EDGE 36$\frac{1}{2}$
6. PURLIN – EDGE 53$\frac{1}{2}$
7.   " – SIDE 56

JACK RAFTERS 16 in. CENTRES DECREASE 21$\frac{1}{2}''$, 18 in.—2′ 0$\frac{1}{4}''$, 24 in.—2′ 8$\frac{1}{4}''$

RUN OF RAFTER ins.	$\frac{1}{2}$	1	2	3	4	5	6	7	8	9	10	11
LENGTH OF RAFTER … …	$\frac{5}{8}$	$1\frac{3}{8}$	$2\frac{3}{4}$	4	$5\frac{3}{8}$	$6\frac{3}{4}$	$8\frac{1}{8}$	$9\frac{3}{8}$	$10\frac{3}{4}$	$12\frac{1}{8}$	$13\frac{1}{2}$	$14\frac{3}{4}$
LENGTH OF HIP … …	$\frac{7}{8}$	$1\frac{5}{8}$	$3\frac{3}{8}$	5	$6\frac{3}{4}$	$8\frac{3}{8}$	10	$11\frac{3}{4}$	$13\frac{3}{8}$	$15\frac{1}{8}$	$16\frac{3}{4}$	$18\frac{3}{8}$

RUN OF RAFTER ft.	1	2	3	4	5	6	7	8	9	10
LENGTH OF RAFTER	$1\cdot4\frac{1}{8}$	$2\cdot8\frac{1}{4}$	$4\cdot0\frac{1}{2}$	$5\cdot4\frac{5}{8}$	$6\cdot8\frac{3}{4}$	$8\cdot0\frac{7}{8}$	$9\cdot5$	$10\cdot9\frac{1}{8}$	$12\cdot1\frac{3}{8}$	$13\cdot5\frac{1}{2}$
LENGTH OF HIP	$1\cdot8\frac{1}{8}$	$3\cdot4\frac{1}{4}$	$5\cdot0\frac{3}{8}$	$6\cdot8\frac{1}{2}$	$8\cdot4\frac{5}{8}$	$10\cdot0\frac{3}{4}$	$11\cdot8\frac{7}{8}$	$13\cdot5$	$15\cdot1$	$16\cdot9\frac{1}{8}$

$42\frac{1}{2}°$ PITCH

RISE OF COMMON RAFTER 11″ PER FOOT OF RUN

BEVELS:
1. COMMON RAFTER – SEAT $42\frac{1}{2}$
2.   ″   ″ – RIDGE $47\frac{1}{2}$
3. HIP OR VALLEY – SEAT 33
4.   ″   ″   ″ – RIDGE 57
5. JACK RAFTER – EDGE $36\frac{1}{4}$
6. PURLIN – EDGE $53\frac{3}{4}$
7.   ″ – SIDE $55\frac{3}{4}$

JACK RAFTERS 16 in. CENTRES DECREASE $21\frac{3}{4}″$, 18 in.—$2′ 0\frac{3}{8}″$, 24 in.—$2′ 8\frac{5}{8}″$

RUN OF RAFTER	ins.	$\frac{1}{2}$	1	2	3	4	5	6	7	8	9	10	11
LENGTH OF RAFTER		$\frac{5}{8}$	$1\frac{3}{8}$	$2\frac{3}{4}$	4	$5\frac{3}{8}$	$6\frac{3}{4}$	$8\frac{1}{8}$	$9\frac{3}{8}$	$10\frac{3}{4}$	$12\frac{1}{4}$	$13\frac{5}{8}$	$14\frac{7}{8}$
LENGTH OF HIP		$\frac{7}{8}$	$1\frac{5}{8}$	$3\frac{3}{8}$	5	$6\frac{3}{4}$	$8\frac{3}{8}$	10	$11\frac{3}{4}$	$13\frac{3}{8}$	$15\frac{1}{8}$	$16\frac{7}{8}$	$18\frac{1}{2}$

RUN OF RAFTER	ft.	1	2	3	4	5	6	7	8	9	10
LENGTH OF RAFTER		$1·4\frac{1}{4}$	$2·8\frac{1}{2}$	$4·0\frac{7}{8}$	$5·5\frac{1}{8}$	$6·9\frac{3}{8}$	$8·1\frac{5}{8}$	9·6	$10·10\frac{1}{4}$	$12·2\frac{1}{2}$	$13·6\frac{3}{4}$
LENGTH OF HIP		$1·8\frac{1}{4}$	$3·4\frac{1}{2}$	$5·0\frac{5}{8}$	6·9	$8·5\frac{1}{8}$	$10·1\frac{1}{8}$	$11·9\frac{5}{8}$	$13·5\frac{7}{8}$	15·2	$16·10\frac{3}{8}$

43° PITCH

RISE OF COMMON RAFTER 11$\frac{3}{16}$″ PER FOOT OF RUN

BEVELS:
1. COMMON RAFTER   – SEAT   43
2.   ″   ″   – RIDGE   47
3. HIP OR VALLEY   – SEAT   33$\frac{1}{2}$
4.   ″   ″   ″   – RIDGE   56$\frac{1}{2}$
5. JACK RAFTER   – EDGE   36
6. PURLIN   – EDGE   54
7.   ″   – SIDE   55$\frac{1}{2}$

JACK RAFTERS 16 in. CENTRES DECREASE 21$\frac{7}{8}$″, 18 in.—2′ 0$\frac{5}{8}$″, 24 in.—2′ 8$\frac{7}{8}$″

RUN OF RAFTER	ins.	$\frac{1}{2}$	1	2	3	4	5	6	7	8	9	10	11
LENGTH OF RAFTER		$\frac{5}{8}$	1$\frac{3}{8}$	2$\frac{3}{4}$	4$\frac{1}{8}$	5$\frac{1}{2}$	6$\frac{7}{8}$	8$\frac{1}{4}$	9$\frac{5}{8}$	10$\frac{7}{8}$	12$\frac{1}{4}$	13$\frac{5}{8}$	15
LENGTH OF HIP		$\frac{7}{8}$	1$\frac{3}{4}$	3$\frac{3}{8}$	5$\frac{1}{8}$	6$\frac{3}{4}$	8$\frac{1}{2}$	10$\frac{1}{8}$	11$\frac{7}{8}$	13$\frac{1}{2}$	15$\frac{1}{4}$	17	18$\frac{5}{8}$

RUN OF RAFTER	ft.	1	2	3	4	5	6	7	8	9	10
LENGTH OF RAFTER		1·4$\frac{3}{8}$	2·8$\frac{7}{8}$	4·1$\frac{1}{4}$	5·5$\frac{5}{8}$	6·10	8·2$\frac{1}{2}$	9·6$\frac{7}{8}$	10·11$\frac{1}{4}$	12·3$\frac{5}{8}$	13·8$\frac{1}{8}$
LENGTH OF HIP		1·8$\frac{3}{8}$	3·4$\frac{5}{8}$	5·1	6·9$\frac{1}{4}$	8·5$\frac{5}{8}$	10·2	11·10$\frac{1}{4}$	13·6$\frac{5}{8}$	15·3	16·11$\frac{1}{4}$

44° PITCH

RISE OF COMMON RAFTER 11$\frac{9}{16}$″ PER FOOT OF RUN

BEVELS:
1. COMMON RAFTER – SEAT 44
2.   ″   ″ – RIDGE 46
3. HIP OR VALLEY – SEAT 34$\frac{1}{2}$
4.   ″   ″   ″ – RIDGE 55$\frac{1}{2}$
5. JACK RAFTER – EDGE 35$\frac{1}{2}$
6. PURLIN – EDGE 54$\frac{1}{2}$
7.   ″ – SIDE 55

JACK RAFTERS 16 in. CENTRES DECREASE 22$\frac{1}{4}$″, 18 in.—2′ 1″, 24 in.—2′ 9$\frac{3}{8}$″

RUN OF RAFTER	ins.	$\frac{1}{2}$	1	2	3	4	5	6	7	8	9	10	11
LENGTH OF RAFTER … …		$\frac{3}{4}$	1$\frac{3}{8}$	2$\frac{3}{4}$	4$\frac{1}{8}$	5$\frac{1}{2}$	7	8$\frac{3}{8}$	9$\frac{3}{4}$	11$\frac{1}{8}$	12$\frac{1}{2}$	13$\frac{7}{8}$	15$\frac{1}{4}$
LENGTH OF HIP … …		$\frac{7}{8}$	1$\frac{3}{4}$	3$\frac{3}{8}$	5$\frac{1}{8}$	6$\frac{7}{8}$	8$\frac{1}{2}$	10$\frac{1}{4}$	12	13$\frac{1}{4}$	15$\frac{3}{8}$	17$\frac{1}{8}$	18$\frac{7}{8}$

RUN OF RAFTER	ft.	1	2	3	4	5	6	7	8	9	10
LENGTH OF RAFTER		1·4$\frac{5}{8}$	2·9$\frac{3}{8}$	4·2$\frac{1}{4}$	5·6$\frac{3}{4}$	6·11$\frac{3}{8}$	8·4$\frac{1}{8}$	9·8$\frac{3}{4}$	11·1$\frac{1}{2}$	12·6$\frac{1}{8}$	13·10$\frac{7}{8}$
LENGTH OF HIP		1·8$\frac{1}{2}$	3·5$\frac{1}{8}$	5·1$\frac{5}{8}$	6·10$\frac{1}{8}$	8·6$\frac{3}{4}$	10·3$\frac{1}{4}$	11·11$\frac{3}{4}$	13·8$\frac{3}{8}$	15·4$\frac{7}{8}$	17·1$\frac{3}{8}$

45° or TRUE PITCH

RISE OF COMMON RAFTER 12″ PER FOOT OF RUN

	BEVELS:	1. COMMON RAFTER	– SEAT	45
		2. ″ ″	– RIDGE	45
		3. HIP OR VALLEY	– SEAT	$35\frac{1}{2}$
		4. ″ ″ ″	– RIDGE	$54\frac{1}{2}$
		5. JACK RAFTER	– EDGE	$35\frac{1}{2}$
		6. PURLIN	– EDGE	$54\frac{1}{2}$
		7. ″	– SIDE	$54\frac{1}{2}$

JACK RAFTERS 16 in. CENTRES DECREASE $22\frac{5}{8}''$, 18 in.—2′ $1\frac{1}{2}''$, 24 in.—2′ 10″

RUN OF RAFTER	ins.	$\frac{1}{2}$	1	2	3	4	5	6	7	8	9	10	11
LENGTH OF RAFTER		$\frac{3}{4}$	$1\frac{3}{8}$	$2\frac{7}{8}$	$4\frac{1}{4}$	$5\frac{5}{8}$	$7\frac{1}{8}$	$8\frac{1}{2}$	$9\frac{7}{8}$	$11\frac{1}{4}$	$12\frac{3}{4}$	$14\frac{1}{8}$	$15\frac{1}{2}$
LENGTH OF HIP		$\frac{7}{8}$	$1\frac{3}{4}$	$3\frac{1}{2}$	$5\frac{1}{4}$	$6\frac{7}{8}$	$8\frac{5}{8}$	$10\frac{3}{8}$	$12\frac{1}{8}$	$13\frac{7}{8}$	$15\frac{5}{8}$	$17\frac{3}{8}$	19

RUN OF RAFTER	ft.	1	2	3	4	5	6	7	8	9	10
LENGTH OF RAFTER		1·5	2·10	$4·2\frac{7}{8}$	$5·7\frac{5}{8}$	$7·0\frac{7}{8}$	$8·5\frac{7}{8}$	$9·10\frac{3}{4}$	$11·3\frac{3}{4}$	$12·8\frac{3}{4}$	$14·1\frac{3}{4}$
LENGTH OF HIP		$1·8\frac{3}{4}$	$3·5\frac{1}{2}$	$5·2\frac{3}{8}$	$6·11\frac{1}{8}$	$8·7\frac{7}{8}$	$10·4\frac{3}{4}$	$12·1\frac{1}{2}$	$13·10\frac{1}{4}$	15·7	$17·3\frac{7}{8}$

46° PITCH

RISE OF COMMON RAFTER 1′ 0$\frac{7}{16}$″ PER FOOT OF RUN

BEVELS :	1.	COMMON RAFTER	– SEAT 46
	2.	″ ″	– RIDGE 44
	3.	HIP OR VALLEY	– SEAT 36
	4.	″ ″ ″	– RIDGE 54
	5.	JACK RAFTER	– EDGE 35
	6.	PURLIN	– EDGE 55
	7.	″	– SIDE 54$\frac{1}{2}$

JACK RAFTERS 16 in. CENTRES DECREASE 23″, 18 in.—2′ 1$\frac{7}{8}$″, 24 in.—2′ 10$\frac{1}{2}$″

RUN OF RAFTER	ins.	$\frac{1}{2}$	1	2	3	4	5	6	7	8	9	10	11
LENGTH OF RAFTER		$\frac{3}{4}$	$1\frac{1}{2}$	$2\frac{7}{8}$	$4\frac{3}{8}$	$5\frac{3}{4}$	$7\frac{1}{4}$	$8\frac{5}{8}$	$10\frac{1}{8}$	$11\frac{1}{2}$	13	$14\frac{3}{8}$	$15\frac{7}{8}$
LENGTH OF HIP		$\frac{7}{8}$	$1\frac{3}{4}$	$3\frac{1}{2}$	$5\frac{1}{4}$	7	$8\frac{3}{4}$	$10\frac{1}{2}$	$12\frac{1}{4}$	14	$15\frac{3}{4}$	$17\frac{1}{2}$	$19\frac{1}{4}$

RUN OF RAFTER	ft.	1	2	3	4	5	6	7	8	9	10
LENGTH OF RAFTER		$1\cdot5\frac{1}{4}$	$2\cdot10\frac{1}{2}$	$4\cdot3\frac{7}{8}$	$5\cdot9\frac{1}{8}$	$7\cdot2\frac{3}{8}$	$8\cdot7\frac{5}{8}$	$10\cdot0\frac{7}{8}$	$11\cdot6\frac{1}{4}$	$12\cdot11\frac{1}{2}$	$14\cdot4\frac{3}{4}$
LENGTH OF HIP		$1\cdot9$	$3\cdot6\frac{1}{8}$	$5\cdot3\frac{1}{8}$	$7\cdot0\frac{1}{8}$	$8\cdot9\frac{1}{8}$	$10\cdot6\frac{1}{4}$	$12\cdot3\frac{1}{4}$	$14\cdot0\frac{1}{4}$	$15\cdot9\frac{1}{4}$	$17\cdot6\frac{3}{8}$

47° PITCH

RISE OF COMMON RAFTER 1′ 0$\frac{7}{8}$″ PER FOOT OF RUN

BEVELS :
1. COMMON RAFTER   – SEAT   47
2.   ″   ″   – RIDGE   43
3. HIP OR VALLEY   – SEAT   37
4.   ″   ″   ″   – RIDGE   53
5. JACK RAFTER   – EDGE   34$\frac{1}{2}$
6. PURLIN   – EDGE   55$\frac{1}{2}$
7.   ″   – SIDE   54

JACK RAFTERS 16 in. CENTRES DECREASE 23$\frac{1}{2}$″, 18 in.—2′ 2$\frac{3}{8}$″, 24 in.—2′ 11$\frac{1}{4}$″

RUN OF RAFTER *ins.*	$\frac{1}{2}$	1	2	3	4	5	6	7	8	9	10	11
LENGTH OF RAFTER … …	$\frac{3}{4}$	$1\frac{1}{2}$	$2\frac{7}{8}$	$4\frac{3}{8}$	$5\frac{7}{8}$	$7\frac{3}{8}$	$8\frac{3}{4}$	$10\frac{1}{4}$	$11\frac{3}{4}$	$13\frac{1}{4}$	$14\frac{5}{8}$	$16\frac{1}{8}$
LENGTH OF HIP … …	$\frac{7}{8}$	$1\frac{3}{4}$	$3\frac{1}{2}$	$5\frac{3}{8}$	$7\frac{1}{8}$	$8\frac{7}{8}$	$10\frac{5}{8}$	$12\frac{3}{8}$	$14\frac{1}{4}$	16	$17\frac{3}{4}$	$19\frac{1}{2}$

RUN OF RAFTER *ft.*	1	2	3	4	5	6	7	8	9	10
LENGTH OF RAFTER	1·5$\frac{5}{8}$	2·11$\frac{1}{4}$	4·4$\frac{3}{4}$	5·10$\frac{3}{8}$	7·4	8·9$\frac{5}{8}$	10·3$\frac{1}{8}$	11·8$\frac{3}{4}$	13·2$\frac{3}{8}$	14·8
LENGTH OF HIP	1·9$\frac{1}{4}$	3·6$\frac{5}{8}$	5·3$\frac{7}{8}$	7·1$\frac{1}{8}$	8·10$\frac{1}{2}$	10·7$\frac{3}{4}$	12·5$\frac{1}{8}$	14·2$\frac{5}{8}$	15·11$\frac{5}{6}$	17·9

48° PITCH

RISE OF COMMON RAFTER 1′ 1$\frac{5}{16}$″ PER FOOT OF RUN

BEVELS:
1. COMMON RAFTER   – SEAT   48
2.   ″   ″   – RIDGE   42
3. HIP OR VALLEY   – SEAT   38
4.   ″   ″   ″   – RIDGE   52
5. JACK RAFTER   – EDGE   34
6. PURLIN   – EDGE   56
7.   ″   – SIDE   53$\frac{1}{2}$

JACK RAFTERS 16 in. CENTRES DECREASE 23$\frac{7}{8}$″, 18 in.—2′ 2$\frac{7}{8}$″, 24 in.—2′ 11$\frac{7}{8}$″

RUN OF RAFTER	ins.	$\frac{1}{2}$	1	2	3	4	5	6	7	8	9	10	11
LENGTH OF RAFTER		$\frac{3}{4}$	1$\frac{1}{2}$	3	4$\frac{1}{2}$	6	7$\frac{1}{2}$	9	10$\frac{1}{2}$	12	13$\frac{1}{2}$	15	16$\frac{1}{2}$
LENGTH OF HIP		$\frac{7}{8}$	1$\frac{3}{4}$	3$\frac{5}{8}$	5$\frac{3}{8}$	7$\frac{1}{4}$	9	10$\frac{3}{4}$	12$\frac{5}{8}$	14$\frac{3}{8}$	16$\frac{1}{8}$	18	19$\frac{3}{4}$

RUN OF RAFTER	ft.	1	2	3	4	5	6	7	8	9	10
LENGTH OF RAFTER		1·5$\frac{7}{8}$	2·11$\frac{7}{8}$	4·5$\frac{3}{4}$	5·11$\frac{3}{4}$	7·5$\frac{5}{8}$	8·11$\frac{5}{8}$	10·5$\frac{1}{2}$	11·11$\frac{1}{2}$	13·5$\frac{3}{8}$	14·11$\frac{3}{8}$
LENGTH OF HIP		1·9$\frac{5}{8}$	3·7$\frac{1}{8}$	5·4$\frac{3}{4}$	7·2$\frac{1}{4}$	8·11$\frac{7}{8}$	10·9$\frac{1}{2}$	12·7	14·4$\frac{5}{8}$	16·2$\frac{1}{8}$	17·11$\frac{3}{4}$

49° PITCH

RISE OF COMMON RAFTER 1′ 1$\frac{13}{16}$″ PER FOOT OF RUN

BEVELS:
1. COMMON RAFTER – SEAT 49
2.   〃   〃   – RIDGE 41
3. HIP OR VALLEY – SEAT 39
4. 〃   〃   〃 – RIDGE 51
5. JACK RAFTER – EDGE 33$\frac{1}{2}$
6. PURLIN – EDGE 56$\frac{1}{2}$
7. 〃 – SIDE 53

JACK RAFTERS 16 in. CENTRES DECREASE 2′ 0$\frac{3}{8}$″, 18 in.—2′ 3$\frac{3}{8}$″, 24 in.—3′ 0$\frac{5}{8}$″

RUN OF RAFTER	ins.	$\frac{1}{2}$	1	2	3	4	5	6	7	8	9	10	11
LENGTH OF RAFTER		$\frac{3}{4}$	1$\frac{1}{2}$	3	4$\frac{5}{8}$	6$\frac{1}{8}$	7$\frac{5}{8}$	9$\frac{1}{8}$	10$\frac{5}{8}$	12$\frac{1}{4}$	13$\frac{3}{4}$	15$\frac{1}{4}$	16$\frac{3}{4}$
LENGTH OF HIP		$\frac{7}{8}$	1$\frac{7}{8}$	3$\frac{5}{8}$	5$\frac{1}{2}$	7$\frac{1}{4}$	9$\frac{1}{8}$	10$\frac{7}{8}$	12$\frac{3}{4}$	14$\frac{5}{8}$	16$\frac{3}{8}$	18$\frac{1}{4}$	20

RUN OF RAFTER	ft.	1	2	3	4	5	6	7	8	9	10
LENGTH OF RAFTER		1·6$\frac{1}{4}$	3·0$\frac{5}{8}$	4·6$\frac{7}{8}$	6·1$\frac{1}{8}$	7·7$\frac{1}{2}$	9·1$\frac{3}{4}$	10·8	12·2$\frac{3}{8}$	13·8$\frac{5}{8}$	15·2$\frac{7}{8}$
LENGTH OF HIP		1·9$\frac{7}{8}$	3·7$\frac{3}{4}$	5·5$\frac{5}{8}$	7·3$\frac{1}{2}$	9·1$\frac{3}{8}$	10·11$\frac{1}{4}$	12·9$\frac{1}{8}$	14·7	16·4$\frac{3}{4}$	18·2$\frac{5}{8}$

50° PITCH

RISE OF COMMON RAFTER 1′ 2$\frac{5}{16}$″ PER FOOT OF RUN

BEVELS:
1. COMMON RAFTER   – SEAT   50
2.   ″   ″   – RIDGE   40
3. HIP OR VALLEY   – SEAT   40
4.   ″   ″   ″   – RIDGE   50
5. JACK RAFTER   – EDGE   32$\frac{1}{2}$
6. PURLIN   – EDGE   57$\frac{1}{2}$
7.   ″   – SIDE   52$\frac{1}{2}$

JACK RAFTERS 16 in. CENTRES DECREASE 2′ 0$\frac{7}{8}$″, 18 in.—2′ 4″, 24 in.—3′ 1$\frac{3}{8}$″

RUN OF RAFTER ins.	$\frac{1}{2}$	1	2	3	4	5	6	7	8	9	10	11
LENGTH OF RAFTER	$\frac{3}{4}$	1$\frac{1}{2}$	3$\frac{1}{8}$	4$\frac{5}{8}$	6$\frac{1}{4}$	7$\frac{3}{4}$	9$\frac{3}{8}$	10$\frac{7}{8}$	12$\frac{1}{2}$	14	15$\frac{1}{2}$	17$\frac{1}{8}$
LENGTH OF HIP	$\frac{7}{8}$	1$\frac{7}{8}$	3$\frac{3}{4}$	5$\frac{1}{2}$	7$\frac{3}{8}$	9$\frac{1}{4}$	11$\frac{1}{8}$	13	14$\frac{3}{4}$	16$\frac{5}{8}$	18$\frac{1}{2}$	20$\frac{3}{8}$

RUN OF RAFTER ft.	1	2	3	4	5	6	7	8	9	10
LENGTH OF RAFTER	1·6$\frac{5}{8}$	3·1$\frac{3}{8}$	4·8	6·2$\frac{5}{8}$	7·9$\frac{3}{8}$	9·4	10·10$\frac{5}{8}$	12·5$\frac{3}{8}$	14·0	15·6$\frac{5}{8}$
LENGTH OF HIP	1·10$\frac{1}{4}$	3·8$\frac{3}{8}$	5·6$\frac{5}{8}$	7·4$\frac{3}{4}$	9·3	11·1$\frac{1}{8}$	12·11$\frac{3}{8}$	14·9$\frac{1}{2}$	16·7$\frac{3}{4}$	18·5$\frac{7}{8}$

51° PITCH

RISE OF COMMON RAFTER 1′ 2$\frac{13}{16}$″ PER FOOT OF RUN

BEVELS :
1. COMMON RAFTER – SEAT 51
2.   ″   ″ – RIDGE 39
3. HIP OR VALLEY – SEAT 41
4. ″ ″   ″ – RIDGE 49
5. JACK RAFTER – EDGE 32
6. PURLIN – EDGE 58
7.   ″ – SIDE 52

JACK RAFTERS 16 in. CENTRES DECREASE 2′ 1$\frac{3}{8}$″, 18 in.—2′ 4$\frac{5}{8}$″, 24 in.—3′ 2$\frac{1}{8}$″

RUN OF RAFTER	ins.	$\frac{1}{2}$	1	2	3	4	5	6	7	8	9	10	11
LENGTH OF RAFTER		$\frac{3}{4}$	$1\frac{5}{8}$	$3\frac{1}{8}$	$4\frac{3}{4}$	$6\frac{3}{8}$	8	$9\frac{1}{2}$	$11\frac{1}{8}$	$12\frac{3}{4}$	$14\frac{1}{4}$	$15\frac{7}{8}$	$17\frac{1}{2}$
LENGTH OF HIP		$\frac{7}{8}$	$1\frac{7}{8}$	$3\frac{3}{4}$	$5\frac{5}{8}$	$7\frac{1}{2}$	$9\frac{3}{8}$	$11\frac{1}{4}$	$13\frac{1}{8}$	15	$16\frac{7}{8}$	$18\frac{3}{4}$	$20\frac{5}{8}$

RUN OF RAFTER	ft.	1	2	3	4	5	6	7	8	9	10
LENGTH OF RAFTER		$1 \cdot 7\frac{1}{8}$	$3 \cdot 2\frac{1}{8}$	$4 \cdot 9\frac{1}{4}$	$6 \cdot 4\frac{1}{4}$	$7 \cdot 11\frac{3}{8}$	$9 \cdot 6\frac{3}{8}$	$11 \cdot 1\frac{1}{2}$	$12 \cdot 8\frac{1}{2}$	$14 \cdot 3\frac{5}{8}$	$15 \cdot 10\frac{5}{8}$
LENGTH OF HIP		$1 \cdot 10\frac{1}{2}$	$3 \cdot 9$	$5 \cdot 7\frac{5}{8}$	$7 \cdot 6\frac{1}{8}$	$9 \cdot 4\frac{5}{8}$	$11 \cdot 3\frac{1}{4}$	$13 \cdot 1\frac{3}{4}$	$15 \cdot 0\frac{1}{4}$	$16 \cdot 10\frac{7}{48}$	$18 \cdot 9\frac{3}{8}$

52° PITCH

RISE OF COMMON RAFTER 1′ 3$\frac{3}{8}$″ PER FOOT OF RUN

BEVELS :
1. COMMON RAFTER — SEAT 52
2.   ″   ″ — RIDGE 38
3. HIP OR VALLEY — SEAT 42
4.   ″   ″   ″ — RIDGE 48
5. JACK RAFTER — EDGE 31$\frac{1}{2}$
6. PURLIN — EDGE 58$\frac{1}{2}$
7.   ″ — SIDE 52

JACK RAFTERS 16 in. CENTRES DECREASE 2′ 2″, 18 in.—2′ 5$\frac{1}{4}$″, 24 in.—3′ 3″

RUN OF RAFTER ins.	$\frac{1}{2}$	1	2	3	4	5	6	7	8	9	10	11
LENGTH OF RAFTER	$\frac{3}{4}$	1$\frac{5}{8}$	3$\frac{1}{4}$	4$\frac{7}{8}$	6$\frac{1}{2}$	8$\frac{1}{8}$	9$\frac{3}{4}$	11$\frac{3}{8}$	13	14$\frac{5}{8}$	16$\frac{1}{4}$	17$\frac{7}{8}$
LENGTH OF HIP	1	1$\frac{7}{8}$	3$\frac{7}{8}$	5$\frac{3}{4}$	7$\frac{5}{8}$	9$\frac{1}{2}$	11$\frac{1}{2}$	13$\frac{3}{8}$	15$\frac{1}{4}$	17$\frac{1}{8}$	19$\frac{1}{8}$	21

RUN OF RAFTER ft.	1	2	3	4	5	6	7	8	9	10
LENGTH OF RAFTER	1·7$\frac{1}{2}$	3·3	4·10$\frac{1}{2}$	6·6	8·1$\frac{1}{2}$	9·9	11·4$\frac{1}{2}$	12·11$\frac{7}{8}$	14·7$\frac{3}{8}$	16·2$\frac{7}{8}$
LENGTH OF HIP	1·10$\frac{7}{8}$	3·9$\frac{3}{4}$	5·8$\frac{5}{8}$	7·7$\frac{1}{2}$	9·6$\frac{3}{8}$	11·5$\frac{1}{4}$	13·4$\frac{1}{4}$	15·3$\frac{1}{8}$	17·2	19·0$\frac{7}{8}$

53° PITCH

RISE OF COMMON RAFTER 1′ 3$\frac{15}{16}$″ PER FOOT OF RUN

BEVELS :
1. COMMON RAFTER   – SEAT   53
2.   ″   ″   – RIDGE   37
3. HIP OR VALLEY   – SEAT   43
4.   ″   ″   ″   – RIDGE   47
5. JACK RAFTER   – EDGE   31
6. PURLIN   – EDGE   59
7.   ″   – SIDE   51$\frac{1}{2}$

JACK RAFTERS 16 in. CENTRES DECREASE 2′ 2$\frac{5}{8}$″, 18 in.—2′ 5$\frac{7}{8}$″, 24 in.—3′ 3$\frac{7}{8}$″

RUN OF RAFTER	*ins.*	$\frac{1}{2}$	1	2	3	4	5	6	7	8	9	10	11
LENGTH OF RAFTER		$\frac{7}{8}$	1$\frac{5}{8}$	3$\frac{3}{8}$	5	6$\frac{5}{8}$	8$\frac{1}{4}$	10	11$\frac{5}{8}$	13$\frac{1}{4}$	15	16$\frac{5}{8}$	18$\frac{1}{4}$
LENGTH OF HIP		1	1$\frac{7}{8}$	3$\frac{7}{8}$	5$\frac{7}{8}$	7$\frac{3}{4}$	9$\frac{3}{4}$	11$\frac{5}{8}$	13$\frac{5}{8}$	15$\frac{1}{2}$	17$\frac{1}{2}$	19$\frac{3}{8}$	21$\frac{3}{8}$

RUN OF RAFTER	*ft.*	1	2	3	4	5	6	7	8	9	10
LENGTH OF RAFTER		1·8	3·3$\frac{7}{8}$	4·11$\frac{7}{8}$	6·7$\frac{3}{4}$	8·3$\frac{3}{4}$	9·11$\frac{5}{8}$	11·7$\frac{5}{8}$	13·3$\frac{1}{2}$	14·11$\frac{1}{2}$	16·7$\frac{3}{8}$
LENGTH OF HIP		1·11$\frac{1}{4}$	3·10$\frac{1}{2}$	5·9$\frac{3}{4}$	7·9$\frac{1}{8}$	9·8$\frac{3}{8}$	11·7$\frac{5}{8}$	13·6$\frac{7}{8}$	15·6$\frac{1}{8}$	17·5$\frac{3}{8}$	19·4$\frac{5}{8}$

54° PITCH

RISE OF COMMON RAFTER 1′ 4½″ PER FOOT OF RUN

BEVELS :
1. COMMON RAFTER   – SEAT   54
2.   ″   ″   – RIDGE   36
3. HIP OR VALLEY   – SEAT   44
4.   ″   ″   ″   – RIDGE   46
5. JACK RAFTER   – EDGE $30\frac{1}{2}$
6. PURLIN   – EDGE $59\frac{1}{2}$
7.   ″   – SIDE   51

JACK RAFTERS 16 in. CENTRES DECREASE 2′ 3″, 18 in.—2′ $6\frac{5}{8}$″, 24 in.—3′ $4\frac{7}{8}$″

RUN OF RAFTER	ins.	$\frac{1}{2}$	1	2	3	4	5	6	7	8	9	10	11
LENGTH OF RAFTER		$\frac{7}{8}$	$1\frac{3}{4}$	$3\frac{3}{8}$	$5\frac{1}{8}$	$6\frac{3}{4}$	$8\frac{1}{2}$	$10\frac{1}{4}$	$11\frac{7}{8}$	$13\frac{5}{8}$	$15\frac{3}{8}$	17	$18\frac{3}{4}$
LENGTH OF HIP		1	2	4	$5\frac{7}{8}$	$7\frac{7}{8}$	$9\frac{7}{8}$	$11\frac{7}{8}$	$13\frac{3}{4}$	$15\frac{3}{4}$	$17\frac{3}{4}$	$19\frac{3}{4}$	$21\frac{3}{4}$

RUN OF RAFTER	ft.	1	2	3	4	5	6	7	8	9	10
LENGTH OF RAFTER		$1 \cdot 8\frac{3}{8}$	$3 \cdot 4\frac{7}{8}$	$5 \cdot 1\frac{1}{4}$	$6 \cdot 9\frac{5}{8}$	$8 \cdot 6\frac{1}{8}$	$10 \cdot 2\frac{1}{2}$	$11 \cdot 10\frac{7}{8}$	$13 \cdot 7\frac{3}{4}$	$15 \cdot 3\frac{3}{4}$	$17 \cdot 0\frac{1}{8}$
LENGTH OF HIP		$1 \cdot 11\frac{5}{8}$	$3 \cdot 11\frac{3}{8}$	$5 \cdot 11$	$7 \cdot 10\frac{3}{4}$	$9 \cdot 10\frac{3}{8}$	$11 \cdot 10\frac{1}{8}$	$13 \cdot 9\frac{3}{4}$	$15 \cdot 9\frac{1}{2}$	$17 \cdot 9\frac{1}{4}$	$19 \cdot 8\frac{7}{8}$

55° PITCH

RISE OF COMMON RAFTER 1′ 5⅛″ PER FOOT OF RUN

BEVELS:
1. COMMON RAFTER   – SEAT   55
2.   ″   ″   – RIDGE   35
3. HIP OR VALLEY   – SEAT $45\frac{1}{2}$
4.   ″   ″   ″   – RIDGE $44\frac{1}{2}$
5. JACK RAFTER   – EDGE   30
6. PURLIN   – EDGE   60
7.   ″   – SIDE $50\frac{1}{2}$

JACK RAFTERS 16 in. CENTRES DECREASE 2′ 3⅞″, 18 in.—2′ 7⅜″, 24 in.—3′ 5⅞″

RUN OF RAFTER *ins.*	$\frac{1}{2}$	1	2	3	4	5	6	7	8	9	10	11
LENGTH OF RAFTER … …	$\frac{7}{8}$	$1\frac{3}{4}$	$3\frac{1}{2}$	$5\frac{1}{4}$	7	$8\frac{3}{4}$	$10\frac{1}{2}$	$12\frac{1}{4}$	14	$15\frac{3}{4}$	$17\frac{3}{8}$	$19\frac{1}{8}$
LENGTH OF HIP … …	1	2	4	6	8	10	12	$14\frac{1}{8}$	$16\frac{1}{8}$	$18\frac{1}{8}$	$20\frac{1}{8}$	$22\frac{1}{8}$

RUN OF RAFTER *ft.*	1	2	3	4	5	6	7	8	9	10
LENGTH OF RAFTER	$1 \cdot 8\frac{7}{8}$	$3 \cdot 5\frac{7}{8}$	$5 \cdot 2\frac{3}{4}$	$6 \cdot 11\frac{5}{8}$	$8 \cdot 8\frac{5}{8}$	$10 \cdot 5\frac{1}{2}$	$12 \cdot 2\frac{1}{2}$	$13 \cdot 11\frac{3}{8}$	$15 \cdot 8\frac{1}{4}$	$17 \cdot 5\frac{1}{4}$
LENGTH OF HIP	$2 \cdot 0\frac{1}{8}$	$4 \cdot 0\frac{1}{4}$	$6 \cdot 0\frac{3}{8}$	$8 \cdot 0\frac{1}{2}$	$10 \cdot 0\frac{5}{8}$	$12 \cdot 0\frac{5}{8}$	$14 \cdot 0\frac{3}{4}$	$16 \cdot 0\frac{7}{8}$	$18 \cdot 1$	$20 \cdot 1\frac{1}{8}$

Note: Italian Pitch (Rise = $\frac{3}{4}$ Span) has a rafter seat bevel of 56° 18'. Lengths, etc., are based on this angle.

ITALIAN PITCH

RISE OF COMMON RAFTER 1′ 6″ PER FOOT OF RUN

BEVELS:
1. COMMON RAFTER – SEAT $56\frac{1}{2}$
2.   ″   ″ – RIDGE $33\frac{1}{2}$
3. HIP OR VALLEY – SEAT $46\frac{1}{2}$
4.   ″   ″   ″ – RIDGE $43\frac{1}{2}$
5. JACK RAFTER – EDGE 29
6. PURLIN – EDGE 61
7.   ″ – SIDE 50

JACK RAFTERS 16 in. CENTRES DECREASE 2′ $4\frac{7}{8}$″, 18 in.—2′ $8\frac{1}{2}$″, 24 in.—3′ $7\frac{1}{4}$″

RUN OF RAFTER *ins.*	$\frac{1}{2}$	1	2	3	4	5	6	7	8	9	10	11
LENGTH OF RAFTER	$\frac{7}{8}$	$1\frac{3}{4}$	$3\frac{5}{8}$	$5\frac{3}{8}$	$7\frac{1}{4}$	9	$10\frac{3}{4}$	$12\frac{5}{8}$	$14\frac{3}{8}$	$16\frac{1}{4}$	18	$19\frac{7}{8}$
LENGTH OF HIP	1	2	$4\frac{1}{8}$	$6\frac{1}{8}$	$8\frac{1}{4}$	$10\frac{1}{4}$	$12\frac{3}{8}$	$14\frac{3}{8}$	$16\frac{1}{2}$	$18\frac{1}{2}$	$20\frac{5}{8}$	$22\frac{5}{8}$

RUN OF RAFTER *ft.*	1	2	3	4	5	6	7	8	9	10
LENGTH OF RAFTER	$1·9\frac{5}{8}$	$3·7\frac{1}{4}$	$5·4\frac{7}{8}$	$7·2\frac{1}{2}$	$9·0\frac{1}{8}$	$10·9\frac{3}{4}$	$12·7\frac{3}{4}$	14·5	$16·2\frac{5}{8}$	$18·0\frac{1}{4}$
LENGTH OF HIP	$2·0\frac{3}{4}$	$4·1\frac{1}{2}$	$6·2\frac{1}{4}$	8·3	$10·3\frac{5}{8}$	$12·4\frac{3}{8}$	$14·5\frac{1}{8}$	$16·5\frac{7}{8}$	$18·6\frac{5}{8}$	$20·7\frac{3}{8}$

57° PITCH

RISE OF COMMON RAFTER 1′ 6$\frac{1}{2}$″ PER FOOT OF RUN

BEVELS :
1. COMMON RAFTER – SEAT 57
2.   ″   ″ – RIDGE 33
3. HIP OR VALLEY – SEAT 47$\frac{1}{2}$
4. ″   ″   ″ – RIDGE 42$\frac{1}{2}$
5. JACK RAFTER – EDGE 28$\frac{1}{2}$
6. PURLIN – EDGE 61$\frac{1}{2}$
7.   ″ – SIDE 50

JACK RAFTERS 16 in. CENTRES DECREASE 2′ 5$\frac{3}{8}$″, 18 in.—2′ 9″, 24 in.—3′ 8$\frac{1}{8}$″

RUN OF RAFTER	ins.	$\frac{1}{2}$	1	2	3	4	5	6	7	8	9	10	11
LENGTH OF RAFTER		$\frac{7}{8}$	$1\frac{7}{8}$	$3\frac{5}{8}$	$5\frac{1}{2}$	$7\frac{3}{8}$	$9\frac{1}{8}$	11	$12\frac{7}{8}$	$14\frac{3}{4}$	$16\frac{1}{2}$	$18\frac{3}{8}$	$20\frac{1}{4}$
LENGTH OF HIP		1	$2\frac{1}{8}$	$4\frac{1}{8}$	$6\frac{1}{4}$	$8\frac{3}{8}$	$10\frac{1}{2}$	$12\frac{1}{2}$	$14\frac{5}{8}$	$16\frac{3}{4}$	$18\frac{3}{4}$	$20\frac{7}{8}$	23

RUN OF RAFTER	ft.	1	2	3	4	5	6	7	8	9	10
LENGTH OF RAFTER		1·10	$3·8\frac{1}{8}$	$5·6\frac{1}{8}$	$7·4\frac{1}{8}$	$9·2\frac{1}{8}$	$11·0\frac{1}{4}$	$12·10\frac{1}{4}$	$14·8\frac{1}{4}$	$16·6\frac{1}{4}$	$18·4\frac{3}{8}$
LENGTH OF HIP		$2·1\frac{1}{8}$	$4·2\frac{1}{8}$	$6·3\frac{1}{4}$	$8·4\frac{3}{8}$	$10·5\frac{3}{8}$	$12·6\frac{1}{2}$	$14·7\frac{5}{8}$	$16·8\frac{3}{4}$	$18·9\frac{3}{4}$	$20·10\frac{7}{8}$

58° PITCH

RISE OF COMMON RAFTER 1′ 7¾″ PER FOOT OF RUN

BEVELS :
1. COMMON RAFTER   – SEAT   58
2.   ″   ″   – RIDGE 32
3. HIP OR VALLEY   – SEAT $48\frac{1}{2}$
4. ″ ″   ″   – RIDGE $41\frac{1}{2}$
5. JACK RAFTER   – EDGE 28
6. PURLIN   – EDGE 62
7.   ″   – SIDE $49\frac{1}{2}$

JACK RAFTERS 16 in. CENTRES DECREASE 2′ 6¼″, 18 in.—2′ 10″, 24 in.—3′ 9¼″

RUN OF RAFTER		*ins.*	$\frac{1}{2}$	1	2	3	4	5	6	7	8	9	10	11
LENGTH OF RAFTER			$\frac{7}{8}$	$1\frac{7}{8}$	$3\frac{3}{4}$	$5\frac{5}{8}$	$7\frac{1}{2}$	$9\frac{3}{8}$	$11\frac{3}{8}$	$13\frac{1}{4}$	$15\frac{1}{8}$	17	$18\frac{7}{8}$	$20\frac{3}{4}$
LENGTH OF HIP			$1\frac{1}{8}$	$2\frac{1}{8}$	$4\frac{1}{4}$	$6\frac{3}{8}$	$8\frac{1}{2}$	$10\frac{5}{8}$	$12\frac{3}{4}$	15	$17\frac{1}{8}$	$19\frac{1}{4}$	$21\frac{3}{8}$	$23\frac{1}{2}$

RUN OF RAFTER	*ft.*	1	2	3	4	5	6	7	8	9	10
LENGTH OF RAFTER		$1 \cdot 10\frac{5}{8}$	$3 \cdot 9\frac{1}{4}$	$5 \cdot 7\frac{7}{8}$	$7 \cdot 6\frac{5}{8}$	$9 \cdot 5\frac{1}{4}$	$11 \cdot 3\frac{7}{8}$	$13 \cdot 2\frac{1}{2}$	$15 \cdot 1\frac{1}{8}$	$16 \cdot 11\frac{3}{4}$	$18 \cdot 10\frac{1}{2}$
LENGTH OF HIP		$2 \cdot 1\frac{1}{8}$	$4 \cdot 3\frac{1}{4}$	$6 \cdot 4\frac{7}{8}$	$8 \cdot 6\frac{1}{2}$	$10 \cdot 8\frac{1}{8}$	$12 \cdot 9\frac{3}{4}$	$14 \cdot 11\frac{3}{8}$	$17 \cdot 1$	$19 \cdot 2\frac{5}{8}$	$21 \cdot 4\frac{1}{4}$

59° PITCH

RISE OF COMMON RAFTER 1′ 8″ PER FOOT OF RUN

BEVELS :
1. COMMON RAFTER   – SEAT   59
2.   ″   ″   – RIDGE   31
3. HIP OR VALLEY   – SEAT $49\frac{1}{2}$
4.   ″   ″   ″   – RIDGE $40\frac{1}{2}$
5. JACK RAFTER   – EDGE $27\frac{1}{2}$
6. PURLIN   – EDGE $62\frac{1}{2}$
7.   ″   – SIDE $49\frac{1}{2}$

JACK RAFTERS 16 in. CENTRES DECREASE 2′ $7\frac{1}{8}$″, 18 in.—2′ 11″, 24 in.—3′ $10\frac{5}{8}$″

RUN OF RAFTER	ins.	$\frac{1}{2}$	1	2	3	4	5	6	7	8	9	10	11
LENGTH OF RAFTER … …		1	2	$3\frac{7}{8}$	$5\frac{7}{8}$	$7\frac{3}{4}$	$9\frac{3}{4}$	$11\frac{5}{8}$	$13\frac{5}{8}$	$15\frac{1}{2}$	$17\frac{1}{2}$	$19\frac{3}{8}$	$21\frac{3}{8}$
LENGTH OF HIP … …		$1\frac{1}{8}$	$2\frac{1}{4}$	$4\frac{3}{8}$	$6\frac{1}{2}$	$8\frac{3}{4}$	$10\frac{7}{8}$	$13\frac{1}{8}$	$15\frac{1}{4}$	$17\frac{1}{2}$	$19\frac{5}{8}$	$21\frac{7}{8}$	24

RUN OF RAFTER ft.	1	2	3	4	5	6	7	8	9	10
LENGTH OF RAFTER	$1·11\frac{1}{4}$	$3·10\frac{5}{8}$	$5·9\frac{7}{8}$	$7·9\frac{1}{4}$	$9·8\frac{1}{2}$	$11·7\frac{3}{4}$	$13·7\frac{1}{8}$	$15·6\frac{3}{8}$	$17·5\frac{3}{4}$	19·5
LENGTH OF HIP	$2·2\frac{1}{4}$	$4·4\frac{3}{8}$	$6·6\frac{5}{8}$	$8·8\frac{7}{8}$	10·11	$13·1\frac{1}{4}$	$15·3\frac{1}{2}$	$17·5\frac{5}{8}$	$19·7\frac{7}{8}$	$21·10\frac{1}{8}$

EQUILATERAL or 60° PITCH

RISE OF COMMON RAFTER 1′ 8$\frac{13}{16}$″ PER FOOT OF RUN

BEVELS:
1. COMMON RAFTER — SEAT 60
2. ″ ″ — RIDGE 30
3. HIP OR VALLEY — SEAT 51
4. ″ ″ ″ — RIDGE 39
5. JACK RAFTER — EDGE 26$\frac{1}{2}$
6. PURLIN — EDGE 63$\frac{1}{2}$
7. ″ — SIDE 49

JACK RAFTERS 16 in. CENTRES DECREASE 2′ 8″, 18 in.—3′ 0″, 24 in.—4′ 0″

RUN OF RAFTER ins.	$\frac{1}{2}$	1	2	3	4	5	6	7	8	9	10	11
LENGTH OF RAFTER … …	1	2	4	6	8	10	12	14	16	18	20	22
LENGTH OF HIP … …	1$\frac{1}{8}$	2$\frac{1}{4}$	4$\frac{1}{2}$	6$\frac{3}{4}$	9	11$\frac{1}{8}$	13$\frac{3}{8}$	15$\frac{5}{8}$	17$\frac{7}{8}$	20$\frac{1}{8}$	22$\frac{3}{8}$	24$\frac{5}{8}$

RUN OF RAFTER ft.	1	2	3	4	5	6	7	8	9	10
LENGTH OF RAFTER	2·0	4·0	6·0	8·0	10·0	12·0	14·0	16·0	18·0	20·0
LENGTH OF HIP	2·2$\frac{7}{8}$	4·5$\frac{5}{8}$	6·8$\frac{1}{2}$	8·11$\frac{1}{4}$	11·2$\frac{1}{8}$	13·5	15·7$\frac{3}{4}$	17·10$\frac{5}{8}$	20·1$\frac{1}{2}$	22·4$\frac{3}{8}$

61° PITCH

RISE OF COMMON RAFTER 1′ 9$\frac{5}{8}$″ PER FOOT OF RUN

BEVELS :
1. COMMON RAFTER   – SEAT   61
2.   ″   ″   – RIDGE 29
3. HIP OR VALLEY   – SEAT   52
4.   ″   ″   ″   – RIDGE 38
5. JACK RAFTER   – EDGE   26
6. PURLIN   – EDGE   64
7.   ″   – SIDE   49

JACK RAFTERS 16 in. CENTRES DECREASE 2′ 9″, 18 in.—3′ 1$\frac{1}{8}$″, 24 in.—4′ 1$\frac{1}{2}$″

RUN OF RAFTER			ins.	$\frac{1}{2}$	1	2	3	4	5	6	7	8	9	10	11
LENGTH OF RAFTER ...	...			1	$2\frac{1}{8}$	$4\frac{1}{8}$	$6\frac{1}{4}$	$8\frac{1}{4}$	$10\frac{3}{8}$	$12\frac{3}{8}$	$14\frac{3}{8}$	$16\frac{1}{2}$	$18\frac{1}{2}$	$20\frac{5}{8}$	$22\frac{3}{4}$
LENGTH OF HIP	...	...		$1\frac{1}{8}$	$2\frac{1}{4}$	$4\frac{5}{8}$	$6\frac{7}{8}$	$9\frac{1}{8}$	$11\frac{1}{2}$	$13\frac{3}{4}$	16	$18\frac{3}{8}$	$20\frac{5}{8}$	$22\frac{7}{8}$	$25\frac{1}{4}$

RUN OF RAFTER	ft.	1	2	3	4	5	6	7	8	9	10
LENGTH OF RAFTER		$2.0\frac{3}{4}$	$4.1\frac{1}{2}$	$6.2\frac{1}{4}$	8.3	$10.3\frac{3}{4}$	$12.4\frac{1}{2}$	$14.5\frac{1}{4}$	16.6	$18.6\frac{3}{4}$	$20.7\frac{1}{2}$
LENGTH OF HIP		$2.3\frac{1}{2}$	4.7	$6.10\frac{1}{2}$	9.2	$11.5\frac{1}{2}$	13.9	$16.0\frac{5}{8}$	$18.4\frac{1}{8}$	$20.7\frac{5}{8}$	$22.11\frac{1}{8}$

62° PITCH

RISE OF COMMON RAFTER 1′ 10$\frac{9}{16}$″ PER FOOT OF RUN

BEVELS :
1. COMMON RAFTER   –   SEAT   62
2.   ″   ″   –   RIDGE   28
3. HIP OR VALLEY   –   SEAT   53
4.   ″   ″   ″   –   RIDGE   37
5. JACK RAFTER   –   EDGE   25
6. PURLIN   –   EDGE   65
7.   ″   –   SIDE   48$\frac{1}{2}$

JACK RAFTERS 16 in. CENTRES DECREASE 2′ 10$\frac{1}{8}$″, 18 in.—3′ 2$\frac{3}{8}$″, 24 in.—4′ 3$\frac{1}{8}$″

RUN OF RAFTER	ins.	$\frac{1}{2}$	1	2	3	4	5	6	7	8	9	10	11
LENGTH OF RAFTER … …		1$\frac{1}{8}$	2$\frac{1}{8}$	4$\frac{1}{4}$	6$\frac{3}{8}$	8$\frac{1}{2}$	10$\frac{5}{8}$	12$\frac{3}{4}$	14$\frac{7}{8}$	17	19$\frac{1}{8}$	21$\frac{1}{4}$	23$\frac{3}{8}$
LENGTH OF HIP … …		1$\frac{1}{8}$	2$\frac{3}{8}$	4$\frac{3}{4}$	7	9$\frac{3}{8}$	11$\frac{3}{4}$	14$\frac{1}{8}$	16$\frac{1}{2}$	18$\frac{7}{8}$	21$\frac{1}{8}$	23$\frac{1}{2}$	25$\frac{7}{8}$

RUN OF RAFTER ft.	1	2	3	4	5	6	7	8	9	10
LENGTH OF RAFTER	2·1$\frac{1}{2}$	4·3$\frac{1}{8}$	6·4$\frac{5}{8}$	8·6$\frac{1}{4}$	10·7$\frac{3}{4}$	12·9$\frac{3}{8}$	14·10$\frac{7}{8}$	17·0$\frac{1}{2}$	19·2	21·3$\frac{5}{8}$
LENGTH OF HIP	2·4$\frac{1}{4}$	4·8$\frac{1}{2}$	7·0$\frac{3}{4}$	9·5	11·9$\frac{1}{8}$	14·1$\frac{3}{8}$	16·5$\frac{5}{8}$	18·9$\frac{7}{8}$	21·2$\frac{1}{8}$	23·6$\frac{3}{8}$

Note: Gothic Pitch (Rise = Span) has a rafter seat bevel of 63° 26'. Lengths, etc., are based on this angle.

GOTHIC PITCH

RISE OF COMMON RAFTER 2′ 0″ PER FOOT OF RUN

BEVELS:
1. COMMON RAFTER – SEAT $63\frac{1}{2}$
2.   ″   ″ – RIDGE $26\frac{1}{2}$
3. HIP OR VALLEY – SEAT $54\frac{1}{2}$
4.   ″   ″   ″ – RIDGE $35\frac{1}{2}$
5. JACK RAFTER – EDGE 24
6. PURLIN – EDGE 66
7.   ″ – SIDE 48

JACK RAFTERS 16 in. CENTRES DECREASE 2′ $11\frac{3}{4}$″, 18 in.—3′ $4\frac{1}{4}$″, 24 in.—4′ $5\frac{5}{8}$″

RUN OF RAFTER	ins.	$\frac{1}{2}$	1	2	3	4	5	6	7	8	9	10	11
LENGTH OF RAFTER		$1\frac{1}{8}$	$2\frac{1}{4}$	$4\frac{1}{2}$	$6\frac{3}{4}$	9	$11\frac{1}{8}$	$13\frac{3}{8}$	$15\frac{5}{8}$	$17\frac{7}{8}$	$20\frac{1}{8}$	$22\frac{3}{8}$	$24\frac{5}{8}$
LENGTH OF HIP		$1\frac{1}{4}$	$2\frac{1}{2}$	$4\frac{7}{8}$	$7\frac{3}{8}$	$9\frac{3}{4}$	$12\frac{1}{4}$	$14\frac{3}{4}$	$17\frac{1}{8}$	$19\frac{5}{8}$	22	$24\frac{1}{2}$	27

RUN OF RAFTER	ft.	1	2	3	4	5	6	7	8	9	10
LENGTH OF RAFTER		$2 \cdot 2\frac{7}{8}$	$4 \cdot 5\frac{5}{8}$	$6 \cdot 8\frac{1}{2}$	$8 \cdot 11\frac{3}{8}$	$11 \cdot 2\frac{1}{8}$	13·5	$15 \cdot 7\frac{7}{8}$	$17 \cdot 10\frac{5}{8}$	$20 \cdot 1\frac{1}{2}$	$22 \cdot 4\frac{1}{4}$
LENGTH OF HIP		$2 \cdot 5\frac{3}{8}$	$4 \cdot 10\frac{3}{4}$	$7 \cdot 4\frac{1}{8}$	$9 \cdot 9\frac{5}{8}$	12·3	$14 \cdot 8\frac{3}{8}$	$17 \cdot 1\frac{3}{4}$	$19 \cdot 7\frac{1}{8}$	$22 \cdot 0\frac{1}{2}$	24·6

65° PITCH

RISE OF COMMON RAFTER 2′ 1¾″ PER FOOT OF RUN

BEVELS:
1. COMMON RAFTER   – SEAT 65
2.   ″   ″   – RIDGE 25
3. HIP OR VALLEY   – SEAT 56½
4.   ″   ″   ″   – RIDGE 33½
5. JACK RAFTER   – EDGE 23
6. PURLIN   – EDGE 67
7.   ″   – SIDE 48

JACK RAFTERS 16 in. CENTRES DECREASE 3′ 1⅞″, 18 in.—3′ 6⅝″, 24 in.—4′ 8¾″

RUN OF RAFTER	ins.	½	1	2	3	4	5	6	7	8	9	10	11
LENGTH OF RAFTER			1⅛	2⅜	4¾	7⅛	9½	11⅞	14¼	16⅝	18⅞	21¼	23⅝	26
LENGTH OF HIP			1¼	2⅝	5⅛	7¾	10¼	12⅞	15⅜	18	20½	23⅛	25¾	28

RUN OF RAFTER	ft.	1	2	3	4	5	6	7	8	9	10
LENGTH OF RAFTER		2·4⅜	4·8¾	7·1⅛	9·5⅝	11·10	14·2⅜	16·6¾	18·11⅛	21·3½	23·8
LENGTH OF HIP		2·6⅞	5·1⅝	7·8½	10·3¼	12·10⅛	15·5	17·11¾	20·6⅝	23·1⅜	25·8¼

66° PITCH

RISE OF COMMON RAFTER 2′ 2$\frac{15}{16}$″ PER FOOT OF RUN

BEVELS:
1. COMMON RAFTER – SEAT 66
2.   ″   ″ – RIDGE 24
3. HIP OR VALLEY – SEAT 58
4.   ″   ″   ″ – RIDGE 32
5. JACK RAFTER – EDGE 22
6. PURLIN – EDGE 68
7.   ″ – SIDE 47$\frac{1}{2}$

JACK RAFTERS 16 in. CENTRES DECREASE 3′ 3$\frac{3}{8}$″, 18 in.—3′ 8$\frac{1}{4}$″, 24 in.—4′ 11″

RUN OF RAFTER	ins.	$\frac{1}{2}$	1	2	3	4	5	6	7	8	9	10	11
LENGTH OF RAFTER		1$\frac{1}{4}$	2$\frac{1}{2}$	4$\frac{7}{8}$	7$\frac{3}{8}$	9$\frac{7}{8}$	12$\frac{1}{4}$	14$\frac{3}{4}$	17$\frac{1}{4}$	19$\frac{5}{8}$	22$\frac{1}{8}$	24$\frac{5}{8}$	27
LENGTH OF HIP		1$\frac{3}{8}$	2$\frac{5}{8}$	5$\frac{1}{4}$	8	10$\frac{5}{8}$	13$\frac{1}{4}$	15$\frac{7}{8}$	18$\frac{5}{8}$	21$\frac{1}{4}$	23$\frac{7}{8}$	26$\frac{1}{2}$	29$\frac{1}{4}$

RUN OF RAFTER	ft.	1	2	3	4	5	6	7	8	9	10
LENGTH OF RAFTER		2·5$\frac{1}{2}$	4·11	7·4$\frac{1}{2}$	9·10	12·3$\frac{1}{2}$	14·9	17·2$\frac{1}{2}$	19·8	22·1$\frac{1}{2}$	24·7
LENGTH OF HIP		2·7$\frac{7}{8}$	5·3$\frac{3}{4}$	7·11$\frac{1}{2}$	10·7$\frac{3}{8}$	13·3$\frac{1}{4}$	15·11	18·6$\frac{7}{8}$	21·2$\frac{3}{4}$	23·10$\frac{5}{8}$	26·6$\frac{3}{8}$

67° PITCH

RISE OF COMMON RAFTER 2′ 4¼″ PER FOOT OF RUN

BEVELS:
1. COMMON RAFTER – SEAT 67
2.   ″   ″   – RIDGE 23
3. HIP OR VALLEY   – SEAT 59
4. ″   ″   ″   – RIDGE 31
5. JACK RAFTER   – EDGE $21\frac{1}{2}$
6. PURLIN   – EDGE $68\frac{1}{2}$
7.   ″   – SIDE $47\frac{1}{2}$

JACK RAFTERS 16 in. CENTRES DECREASE 3′ 5″, 18 in.—3′ $10\frac{1}{8}$″, 24 in.—5′ $1\frac{1}{2}$″

RUN OF RAFTER				ins.	$\frac{1}{2}$	1	2	3	4	5	6	7	8	9	10	11
LENGTH OF RAFTER	...	...			$1\frac{1}{4}$	$2\frac{1}{2}$	$5\frac{1}{8}$	$7\frac{5}{8}$	$10\frac{1}{4}$	$12\frac{3}{4}$	$15\frac{3}{8}$	$17\frac{7}{8}$	$20\frac{1}{2}$	23	$25\frac{5}{8}$	$28\frac{1}{8}$
LENGTH OF HIP	...	...			$1\frac{3}{8}$	$2\frac{3}{4}$	$5\frac{1}{2}$	$8\frac{1}{4}$	11	$13\frac{3}{4}$	$16\frac{1}{2}$	$19\frac{1}{4}$	22	$24\frac{3}{4}$	$27\frac{1}{2}$	$30\frac{1}{4}$

RUN OF RAFTER	ft.	1	2	3	4	5	6	7	8	9	10
LENGTH OF RAFTER		$2 \cdot 6\frac{3}{4}$	$5 \cdot 1\frac{3}{8}$	$7 \cdot 8\frac{1}{8}$	$10 \cdot 2\frac{7}{8}$	$12 \cdot 9\frac{1}{2}$	$15 \cdot 4\frac{1}{4}$	$17 \cdot 11$	$20 \cdot 5\frac{3}{4}$	$23 \cdot 0\frac{3}{8}$	$25 \cdot 7\frac{1}{8}$
LENGTH OF HIP		$2 \cdot 9$	$5 \cdot 5\frac{7}{8}$	$8 \cdot 2\frac{7}{8}$	$10 \cdot 11\frac{7}{8}$	$13 \cdot 8\frac{7}{8}$	$16 \cdot 5\frac{3}{4}$	$19 \cdot 2\frac{3}{4}$	$21 \cdot 11\frac{3}{4}$	$24 \cdot 8\frac{3}{4}$	$27 \cdot 5\frac{5}{8}$

68° PITCH

RISE OF COMMON RAFTER 2′ 5$\frac{11}{16}$″ PER FOOT OF RUN

BEVELS :
1. COMMON RAFTER — SEAT 68
2.   ″   ″ — RIDGE 22
3. HIP OR VALLEY — SEAT 60$\frac{1}{2}$
4. ″   ″   ″ — RIDGE 29$\frac{1}{2}$
5. JACK RAFTER — EDGE 20$\frac{1}{2}$
6. PURLIN — EDGE 69$\frac{1}{2}$
7.   ″ — SIDE 47

JACK RAFTERS 16 in. CENTRES DECREASE 3′ 6$\frac{3}{4}$″, 18 in.—4′ 0″, 24 in.—5′ 4$\frac{1}{8}$″

RUN OF RAFTER ins.	$\frac{1}{2}$	1	2	3	4	5	6	7	8	9	10	11
LENGTH OF RAFTER	1$\frac{3}{8}$	2$\frac{5}{8}$	5$\frac{3}{8}$	8	10$\frac{5}{8}$	13$\frac{3}{8}$	16	18$\frac{5}{8}$	21$\frac{3}{8}$	24	26$\frac{3}{4}$	29$\frac{3}{8}$
LENGTH OF HIP	1$\frac{3}{8}$	2$\frac{7}{8}$	5$\frac{3}{4}$	8$\frac{1}{2}$	11$\frac{3}{8}$	14$\frac{1}{4}$	17$\frac{1}{8}$	20	22$\frac{3}{4}$	25$\frac{5}{8}$	28$\frac{1}{2}$	31$\frac{3}{8}$

RUN OF RAFTER ft.	1	2	3	4	5	6	7	8	9	10
LENGTH OF RAFTER	2·8	5·4$\frac{1}{8}$	8·0$\frac{1}{8}$	10·8$\frac{1}{8}$	13·4$\frac{1}{8}$	16·0$\frac{1}{4}$	18·8$\frac{1}{4}$	21·4$\frac{1}{4}$	24·0$\frac{1}{4}$	26·8$\frac{3}{8}$
LENGTH OF HIP	2·10$\frac{1}{4}$	5·8$\frac{3}{8}$	8·6$\frac{5}{8}$	11·4$\frac{7}{8}$	14·3	17·1$\frac{1}{4}$	19·11$\frac{1}{2}$	22·9$\frac{5}{8}$	25·7$\frac{7}{8}$	28·6$\frac{1}{8}$

69° PITCH

RISE OF COMMON RAFTER 2′ 7¼″ PER FOOT OF RUN

BEVELS :
1. COMMON RAFTER – SEAT 69
2.   ″   ″ – RIDGE 21
3. HIP OR VALLEY – SEAT $61\frac{1}{2}$
4.   ″   ″   ″ – RIDGE $28\frac{1}{2}$
5. JACK RAFTER – EDGE $19\frac{1}{2}$
6. PURLIN – EDGE $70\frac{1}{2}$
7.   ″ – SIDE 47

JACK RAFTERS 16 in. CENTRES DECREASE 3′ $8\frac{5}{8}$″, 18 in.—4′ $2\frac{1}{4}$″, 24 in.—5′ 7″

RUN OF RAFTER	*ins.*	$\frac{1}{2}$	1	2	3	4	5	6	7	8	9	10	11
LENGTH OF RAFTER		$1\frac{3}{8}$	$2\frac{3}{4}$	$5\frac{3}{8}$	$8\frac{3}{8}$	$11\frac{1}{8}$	14	$16\frac{3}{4}$	$19\frac{1}{2}$	$22\frac{3}{8}$	$25\frac{1}{8}$	$27\frac{7}{8}$	$30\frac{3}{4}$
LENGTH OF HIP		$1\frac{1}{2}$	3	$5\frac{7}{8}$	$8\frac{7}{8}$	$11\frac{7}{8}$	$14\frac{7}{8}$	$17\frac{3}{4}$	$20\frac{3}{4}$	$23\frac{3}{4}$	$26\frac{5}{8}$	$29\frac{5}{8}$	$32\frac{5}{8}$

RUN OF RAFTER	*ft.*	1	2	3	4	5	6	7	8	9	10
LENGTH OF RAFTER		$2\cdot9\frac{1}{2}$	$5\cdot7$	$8\cdot4\frac{1}{4}$	$11\cdot2$	$13\cdot11\frac{3}{8}$	$16\cdot8\frac{7}{8}$	$19\cdot6\frac{3}{8}$	$22\cdot3\frac{7}{8}$	$25\cdot1\frac{3}{8}$	$27\cdot10\frac{7}{8}$
LENGTH OF HIP		$2\cdot11\frac{5}{8}$	$5\cdot11\frac{1}{8}$	$8\cdot10\frac{3}{4}$	$11\cdot10\frac{1}{4}$	$14\cdot9\frac{7}{8}$	$17\cdot9\frac{3}{8}$	$20\cdot9$	$23\cdot8\frac{1}{2}$	$26\cdot8\frac{1}{8}$	$29\cdot7\frac{5}{8}$

70° PITCH

RISE OF COMMON RAFTER 2′ 9″ PER FOOT OF RUN

BEVELS :
1. COMMON RAFTER – SEAT 70
2.   ″   ″   – RIDGE 20
3. HIP OR VALLEY – SEAT 63
4.   ″   ″   ″   – RIDGE 27
5. JACK RAFTER – EDGE 19
6. PURLIN   – EDGE 71
7.   ″   – SIDE   47

JACK RAFTERS 16 in. CENTRES DECREASE 3′ $10\frac{3}{4}$″, 18 in.—4′ $4\frac{5}{8}$″, 24 in.—5′ $10\frac{1}{8}$″

RUN OF RAFTER	ins.	$\frac{1}{2}$	1	2	3	4	5	6	7	8	9	10	11
LENGTH OF RAFTER … …		$1\frac{1}{2}$	$2\frac{7}{8}$	$5\frac{7}{8}$	$8\frac{3}{4}$	$11\frac{3}{4}$	$14\frac{5}{8}$	$17\frac{1}{2}$	$20\frac{1}{2}$	$23\frac{3}{8}$	$26\frac{3}{8}$	$29\frac{1}{4}$	$32\frac{1}{8}$
LENGTH OF HIP … …		$1\frac{1}{2}$	$3\frac{1}{8}$	$6\frac{1}{8}$	$9\frac{1}{4}$	$12\frac{3}{8}$	$15\frac{1}{2}$	$18\frac{1}{2}$	$21\frac{5}{8}$	$24\frac{3}{4}$	$27\frac{3}{4}$	$30\frac{7}{8}$	34

RUN OF RAFTER	ft.	1	2	3	4	5	6	7	8	9	10
LENGTH OF RAFTER		$2 \cdot 11\frac{1}{8}$	$5 \cdot 10\frac{1}{8}$	$8 \cdot 9\frac{1}{4}$	$11 \cdot 8\frac{3}{8}$	$14 \cdot 7\frac{3}{8}$	$17 \cdot 6\frac{1}{2}$	$20 \cdot 5\frac{5}{8}$	$23 \cdot 4\frac{5}{8}$	$26 \cdot 3\frac{3}{4}$	$29 \cdot 2\frac{7}{8}$
LENGTH OF HIP		$3 \cdot 1\frac{1}{8}$	$6 \cdot 2\frac{1}{8}$	$9 \cdot 3\frac{1}{4}$	$12 \cdot 4\frac{3}{8}$	$15 \cdot 5\frac{3}{8}$	$18 \cdot 6\frac{1}{2}$	$21 \cdot 7\frac{1}{2}$	$24 \cdot 8\frac{5}{8}$	$27 \cdot 9\frac{3}{4}$	$30 \cdot 10\frac{3}{4}$

71° PITCH

RISE OF COMMON RAFTER 2′ 10$\frac{7}{8}$″ PER FOOT OF RUN

BEVELS :　1. COMMON RAFTER　– SEAT　71
　　　　　2. 　　″　　　　″　　– RIDGE　19
　　　　　3. HIP OR VALLEY　– SEAT　64
　　　　　4. 　″　　″　　　″　　– RIDGE　26
　　　　　5. JACK RAFTER　　– EDGE　18
　　　　　6. PURLIN　　　　　– EDGE　72
　　　　　7. 　　″　　　　　　– SIDE　46$\frac{1}{2}$

JACK RAFTERS 16 in. CENTRES DECREASE 4′ 1$\frac{1}{8}$″, 18 in.—4′ 7$\frac{1}{4}$″, 24 in.—6′ 1$\frac{3}{4}$″

RUN OF RAFTER	*ins.*	$\frac{1}{2}$	1	2	3	4	5	6	7	8	9	10	11
LENGTH OF RAFTER … …		1$\frac{1}{2}$	3$\frac{1}{8}$	6$\frac{1}{8}$	9$\frac{1}{4}$	12$\frac{1}{4}$	15$\frac{3}{8}$	18$\frac{3}{8}$	21$\frac{1}{2}$	24$\frac{5}{8}$	27$\frac{5}{8}$	30$\frac{3}{4}$	33$\frac{3}{4}$
LENGTH OF HIP … …		1$\frac{5}{8}$	3$\frac{1}{4}$	6$\frac{1}{2}$	9$\frac{3}{4}$	12$\frac{7}{8}$	16$\frac{1}{8}$	19$\frac{3}{8}$	22$\frac{5}{8}$	25$\frac{7}{8}$	29$\frac{1}{8}$	32$\frac{1}{4}$	35$\frac{1}{2}$

RUN OF RAFTER	*ft.*	1	2	3	4	5	6	7	8	9	10
LENGTH OF RAFTER		3·0$\frac{7}{8}$	6·1$\frac{3}{4}$	9·2$\frac{5}{8}$	12·3$\frac{3}{8}$	15·4$\frac{1}{4}$	18·5$\frac{1}{8}$	21·6	24·6$\frac{7}{8}$	27·7$\frac{3}{4}$	30·8$\frac{5}{8}$
LENGTH OF HIP		3·2$\frac{3}{4}$	6·5$\frac{1}{2}$	9·8$\frac{1}{4}$	12·11	16·1$\frac{3}{4}$	19·4$\frac{1}{2}$	22·7$\frac{3}{8}$	25·10$\frac{1}{8}$	29·0$\frac{7}{8}$	32·3$\frac{5}{8}$

72° PITCH

RISE OF COMMON RAFTER 3′ 0$\frac{15}{16}$″ PER FOOT OF RUN

BEVELS :
1. COMMON RAFTER – SEAT   72
2.   ″   ″   – RIDGE 18
3. HIP OR VALLEY – SEAT   65$\frac{1}{2}$
4.   ″   ″   ″   – RIDGE 24$\frac{1}{2}$
5. JACK RAFTER – EDGE 17
6. PURLIN – EDGE 73
7.   ″ – SIDE   46$\frac{1}{2}$

JACK RAFTERS 16 in. CENTRES DECREASE 4′ 3$\frac{3}{4}$″, 18 in.—4′ 10$\frac{1}{4}$″, 24 in.—6′ 5$\frac{5}{8}$″

RUN OF RAFTER	*ins.*	$\frac{1}{2}$	1	2	3	4	5	6	7	8	9	10	11
LENGTH OF RAFTER		1$\frac{5}{8}$	3$\frac{1}{4}$	6$\frac{1}{2}$	9$\frac{3}{4}$	13	16$\frac{1}{8}$	19$\frac{3}{8}$	22$\frac{5}{8}$	25$\frac{7}{8}$	29$\frac{1}{8}$	32$\frac{3}{8}$	35$\frac{5}{8}$
LENGTH OF HIP		1$\frac{3}{4}$	3$\frac{3}{8}$	6$\frac{3}{4}$	10$\frac{1}{8}$	13$\frac{1}{2}$	16$\frac{7}{8}$	20$\frac{3}{8}$	23$\frac{3}{4}$	27$\frac{1}{8}$	30$\frac{1}{2}$	33$\frac{7}{8}$	37$\frac{1}{4}$

RUN OF RAFTER	*ft.*	1	2	3	4	5	6	7	8	9	10
LENGTH OF RAFTER		3·2$\frac{7}{8}$	6·5$\frac{5}{8}$	9·8$\frac{1}{2}$	12·11$\frac{3}{8}$	16·2$\frac{1}{8}$	19·5	22·7$\frac{7}{8}$	25·10$\frac{5}{8}$	29·1$\frac{1}{2}$	32·4$\frac{3}{8}$
LENGTH OF HIP		3·4$\frac{5}{8}$	6·9$\frac{1}{4}$	10·1$\frac{7}{8}$	13·6$\frac{1}{2}$	16·11$\frac{1}{8}$	20·3$\frac{3}{4}$	23·8$\frac{3}{8}$	27·1	30·5$\frac{3}{4}$	33·10$\frac{3}{8}$

73° PITCH

RISE OF COMMON RAFTER 3′ 3¼″ PER FOOT OF RUN

BEVELS :
1. COMMON RAFTER   – SEAT   73
2.   ″   ″   – RIDGE 17
3. HIP OR VALLEY   – SEAT $66\frac{1}{2}$
4.   ″   ″   ″ – RIDGE $23\frac{1}{2}$
5. JACK RAFTER   – EDGE $16\frac{1}{2}$
6. PURLIN   – EDGE $73\frac{1}{2}$
7.   ″   – SIDE $46\frac{1}{2}$

JACK RAFTERS 16 in. CENTRES DECREASE 4′ $6\frac{3}{4}$″, 18 in.—5′ $1\frac{1}{2}$″, 24 in.—6′ $10\frac{1}{8}$″

RUN OF RAFTER	ins.	$\frac{1}{2}$	1	2	3	4	5	6	7	8	9	10	11
LENGTH OF RAFTER … …		$1\frac{3}{4}$	$3\frac{3}{8}$	$6\frac{7}{8}$	$10\frac{1}{4}$	$13\frac{5}{8}$	$17\frac{1}{8}$	$20\frac{1}{2}$	24	$27\frac{3}{8}$	$30\frac{3}{4}$	$34\frac{1}{4}$	$37\frac{5}{8}$
LENGTH OF HIP … …		$1\frac{3}{4}$	$3\frac{5}{8}$	$7\frac{1}{8}$	$10\frac{3}{4}$	$14\frac{1}{4}$	$17\frac{7}{8}$	$21\frac{3}{8}$	25	$28\frac{1}{8}$	$32\frac{1}{8}$	$35\frac{5}{8}$	$39\frac{1}{4}$

RUN OF RAFTER	ft.	1	2	3	4	5	6	7	8	9	10
LENGTH OF RAFTER		3·5	$6·10\frac{1}{8}$	$10·3\frac{3}{8}$	$13·8\frac{3}{8}$	$17·1\frac{1}{4}$	$20·6\frac{1}{4}$	$23·11\frac{1}{4}$	$27·4\frac{3}{8}$	$30·9\frac{3}{8}$	$34·2\frac{3}{8}$
LENGTH OF HIP		$3·6\frac{3}{4}$	$7·1\frac{1}{2}$	$10·8\frac{1}{4}$	14·3	$17·9\frac{3}{4}$	$21·4\frac{1}{2}$	$24·11\frac{1}{4}$	28·6	$32·0\frac{3}{4}$	$35·7\frac{1}{2}$

74° PITCH

RISE OF COMMON RAFTER 3′ 5$\frac{7}{8}$″ PER FOOT OF RUN

BEVELS :
1. COMMON RAFTER – SEAT 74
2.   ″   ″ – RIDGE 16
3. HIP OR VALLEY – SEAT 68
4.   ″   ″   ″ – RIDGE 22
5. JACK RAFTER – EDGE 15$\frac{1}{2}$
6. PURLIN – EDGE 74$\frac{1}{2}$
7.   ″ – SIDE 46

JACK RAFTERS 16 in. CENTRES DECREASE 4′ 10″, 18 in.—5′ 5$\frac{1}{4}$″, 24 in.—7′ 3$\frac{1}{8}$″

RUN OF RAFTER *ins.*	$\frac{1}{2}$	1	2	3	4	5	6	7	8	9	10	11
LENGTH OF RAFTER	1$\frac{7}{8}$	3$\frac{5}{8}$	7$\frac{1}{4}$	10$\frac{7}{8}$	14$\frac{1}{2}$	18$\frac{1}{8}$	21$\frac{3}{4}$	25$\frac{3}{8}$	29	32$\frac{5}{8}$	36$\frac{1}{4}$	39$\frac{7}{8}$
LENGTH OF HIP	1$\frac{7}{8}$	3$\frac{3}{4}$	7$\frac{1}{2}$	11$\frac{1}{4}$	15	18$\frac{3}{4}$	22$\frac{5}{8}$	26$\frac{3}{8}$	30$\frac{1}{8}$	33$\frac{5}{8}$	37$\frac{5}{8}$	41$\frac{3}{8}$

RUN OF RAFTER *ft.*	1	2	3	4	5	6	7	8	9	10
LENGTH OF RAFTER	3·7$\frac{1}{2}$	7·3$\frac{1}{8}$	10·10$\frac{5}{8}$	14·6$\frac{1}{8}$	18·1$\frac{5}{8}$	21·9$\frac{1}{4}$				
LENGTH OF HIP	3·9$\frac{1}{8}$	7·6$\frac{1}{4}$	11·3$\frac{1}{2}$	15·0$\frac{5}{8}$	18·9$\frac{3}{4}$	22·7				

75° PITCH

RISE OF COMMON RAFTER 3′ 8$\frac{13}{16}$″ PER FOOT OF RUN

BEVELS:	1. COMMON RAFTER	– SEAT	75
	2. ″ ″	– RIDGE	15
	3. HIP OR VALLEY	– SEAT	69
	4. ″ ″ ″	– RIDGE	21
	5. JACK RAFTER	– EDGE	14$\frac{1}{2}$
	6. PURLIN	– EDGE	75$\frac{1}{2}$
	7. ″	– SIDE	46

Note—Pitches steeper than 75° are not likely to be met with by the craftsman except in steeples, spires and turrets. These call for special construction and are not within the scope of this ready reckoner.

JACK RAFTERS 16 in. CENTRES DECREASE 5′ 1$\frac{7}{8}$″, 18 in.—5′ 9$\frac{1}{2}$″

RUN OF RAFTER	ins.	$\frac{1}{2}$	1	2	3	4	5	6	7	8	9	10	11
LENGTH OF RAFTER		1$\frac{7}{8}$	3$\frac{7}{8}$	7$\frac{3}{4}$	11$\frac{5}{8}$	15$\frac{1}{2}$	19$\frac{3}{8}$	23$\frac{1}{8}$	27	30$\frac{7}{8}$	34$\frac{3}{4}$	38$\frac{5}{8}$	42$\frac{1}{2}$
LENGTH OF HIP		2	4	8	12	16	20	24	27$\frac{7}{8}$	31$\frac{7}{8}$	35$\frac{7}{8}$	39$\frac{7}{8}$	43$\frac{7}{8}$

RUN OF RAFTER	ft.	1	2	3	4	5	6	7	8	9	10
LENGTH OF RAFTER		3·10$\frac{3}{8}$	7·8$\frac{3}{4}$	11·7$\frac{1}{8}$	15·5$\frac{1}{2}$	19·3$\frac{7}{8}$	23·2$\frac{1}{8}$				
LENGTH OF HIP		3·11$\frac{7}{8}$	7·11$\frac{3}{4}$	11·11$\frac{5}{8}$	15·11$\frac{1}{2}$	19·11$\frac{3}{8}$	23·11$\frac{3}{8}$				

6 WALL PLATE AND GABLE STRAPPING

Earlier in this book it was stated that the wall plate is the foundation to the roof, and like all foundations needs to be sound and secure. The wall plate should be strapped down to the building structure; this is usually done by using steel straps as illustrated in Fig. 12(a). The straps can either be built into the brick wall below, or face-fixed by screwing into the wall. On timber framed housing the plate may be adequately secured to the frame by nailing at centres specified by the designer.

Gable walls, especially those on steep pitched roof constructions where the gable is very tall, rely on the roof for their stability and NOT the other way around. Wind blowing on one gable exerts a pressure on it pushing it into the roof, whilst at the opposite end it creates a suction which attempts to suck the gable from the roof. It is therefore a requirement of the building regulations that the gables must be adequately tied back into the roof to give them support. Fig. 12(b) illustrates a typical gable end restraint system on a trussed rafter roof, but this equally applies to traditional roof construction. Building the purlins into the gable will help, but will only be effective if the purlin is mechanically fixed to the wall with some additional form of cleat or strap. Straps are normally placed at approximately 2 m centres, and it is essential that the strap is supported by solid blocking beneath it to ensure that it does not buckle, and that the last rafter is solidly blocked to the gable wall itself.

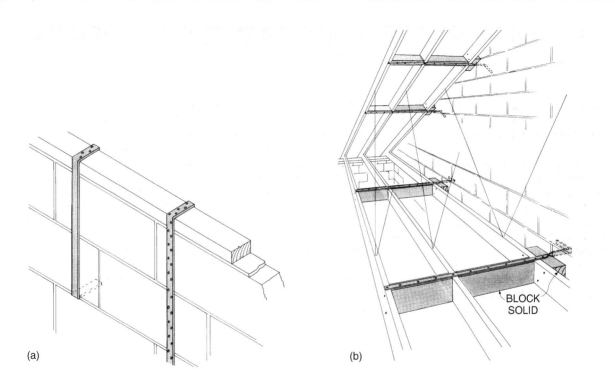

(a)

(b)

BLOCK
SOLID

Fig. 12

7 WIND BRACING

From the previous chapter it can be seen that wind plays a considerable part in destabilising a structure and measures must be taken to ensure the stability of the roof construction in high wind situations. From the previous chapter strapping the gables to the roof have ensured their integrity with the roof, but apart from the binders, purlins and ridge, which connect the rafter members on a horizontal plane, the roof structure is still no more than a number of vertical members of timber connected with a limited number of nails to the members mentioned above. The roof is in effect no more than a set of dominoes standing vertically on their ends and which can be easily made to fall if a pressure is applied to the last one in line, in the case of a roof this would be the gable end. To prevent the domino toppling effect, wind bracing is introduced into the structure to triangulate it on a vertical plane. The elements being discussed here are set out in Fig. 13, which details the bracing required for a typical trussed rafter roof, much of this being applicable to a traditionally constructed roof without hips. Even on a hip roof where the ridge is twice as long as the length on plan of the hip itself it is wise to introduce wind bracing.

Wind bracing is usually timber typically of 25×100 mm ($1'' \times 4''$) in cross section, fitted from wall plate to the ridge at an angle of approximately $45°$ on the underside of the rafters. At each rafter crossing the wind bracing should be nailed to the rafter with 3 nails. The braces should be fitted from the foot of the gable to the ridge on both sides of the roof and then at $45°$ back down again to the plate over the entire length of the roof. This is brace F in Fig. 13; brace H may well be replaced by a purlin in a traditionally cut roof, and brace G by the ridge board. Brace G at ceiling joist level, may well be ceiling joist binders on a traditionally cut roof, whilst K should be fitted to all types of roof construction. Brace J applies only to trussed rafter roof construction.

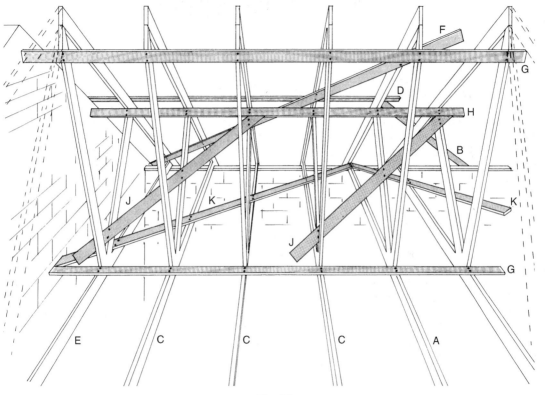

Fig. 13

8 ROOFING METALWORK

The use of smaller timber sections particularly in engineered roofs such as trussed rafter prefabricated assemblies, has led to the increased use of metalwork to join the various members together. Traditional 'tosh' or 'skew' nailing can easily split these smaller timber sections resulting in a poor connection. Fig. 14 illustrates some items of roofing metalwork.

All metalwork used in roof construction should be galvanised; this includes the nails to prevent corrosion. The modern roof can suffer from localised condensation especially if not adequately ventilated and this could lead to premature corrosion of nails and fixings. Whilst a galvanised nail has a slightly rough surface compared to a plain wire nail and therefore gives an improved resistance to movement in the joint, many of the metal to timber connections are specified to be fixed with square twisted galvanised nails which give a far improved performance in the joint.

Typical roofing metalwork would be as follows:

(a) Wall plate straps cross section 30 mm × 1.5/2 mm
(b) Gable restraint straps 30 mm × 5 mm.
(c) Trussed rafter clips to hold truss or rafter to wall plate.
(d) Hip corner tie to hold hip to wall plate at corner.
(e) Girder truss shoes to carry trussed rafters on girder truss support points.
(f) Multi nail plates for coupling timbers into longer lengths.
(g) Framing anchors to connect various trimmed openings and elements in the ceiling joist structure.

Care must be taken to use the nails specified by the metalwork manufacturer both in the type of nail to be used and the number of nails to be used in each joint.

TCP

40

40

138

60

H2.5

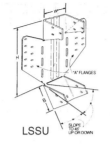

"A" FLANGES

LSSU

SLOPE TO 45° UP OR DOWN

Typical LSU 26
Installation

Typical LSSU
Installation

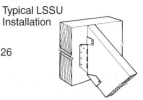

3 Attach hanger to the carrying
member, acute angle side first.
Install nails at an angle.

Typical TCP Installation

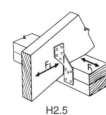

H2.5
Ties rafter
to two plates

A FLANGE   W   A FLANGE

H

SLOPE UP OR DOWN TO 45°

LSU 26

3.75 x 32 mm
nails

1 Nail hanger to slope-cut
carried member, installing seat
nail first. No bevel necessary for
skewed installation.

2 Skew flange to form acute
angle. Bend other flange
back along centreline of
slots. Bend one time only.

ACUTE
ANGLE

Fig. 14
181

9 TOOLS AND EQUIPMENT

The roofing carpenter will need a number of tools and pieces of equipment to satisfactorily obtain information from drawings, mark the timber, set out the roof on the wall plate, cut the timber, and check the completed roof for line, level and plumb. A conventional pencil or pen may be needed for paper calculations, but the true carpenter's pencil should be used for marking timber.

Obtaining Information from the Drawing

(a)   Scale rule: only to be used if dimensions are not clearly shown on the drawing.
(b)   Protractor: to measure the angle of the roof, but again only if the angle is not written on the drawing.
(c)   To mark out the length and angles to be cut on the timbers: steel measuring tape of minimum 5 m in length now available with a digital display to remove the possible error of misreading.
(d)   A bevel: a simple carpenter's adjustable bevel is adequate, this being set to the protractor to obtain cutting angles.
(e)   Alternatively a combination square with centre head and built-in protractor is more versatile.
(f)   Alternatively a digital bevel may be used for instant visual display of the setting angles required.
(g)   A traditional roofing square can be used if the carpenter is trained in the use of this particular tool.

To Cut the Roof

(h)   Hand saw.
(i)   Mains or 110 volt electric hand saw. Cordless powered hand saws are available, but a mains power source is still required to recharge batteries.

(j)   A compound angle mitre saw. This is an electrically powered saw designed especially to cut angles on timbers – check that the saw is large enough to cope with the length of the cut required, a 300 mm diameter saw should be adequate. This type of saw can tilt in both planes and therefore be set to cut compound angles i.e. the ridge and edge bevels on hip jack rafters, in one operation. This type of saw is invariably fixed to a bench or stand, and therefore support will be needed for long timbers to be cut. This support should be fitted with a sliding stop system to allow quick repeat lengths to be cut without remeasuring. NB for safety all power tools on site should be 110 volts – for home use a 240 volt saw may be used only in conjunction with a power protection plug adapter – this will protect against accidentally cut cables and faulty wiring possibly causing electric shock to the operator.

Setting-Up the Roof Structure

(k)   A steel tape at least as long as the roof wall plate itself.
(l)   A good level, by this we mean a good quality level at least 900 mm long with two spirit levels one for horizontal use and one for vertical use for plumbing timbers.
(m) Alternatively a good quality level as above but with digital angle readout would be useful to check work on the roof as it proceeds.
(n)   To check and set level and verticals over longer distances (i.e. beyond the 900 mm of the levels itself), a laser level could be used. A wide range of laser beam projecting levels are now available giving an accurate beam projection of up to 50 m. The sophisticated rotary laser levels will give both horizontal and vertical beam projection, but it must be remembered that these are accurate only if they are mounted on a stable structure.

Learning Resources
Centre